Advance Praise for *BioDiet*

BioDiet is a masterful combination of both the background science and how-to of the ketogenic diet. It is a perfect entrée for someone starting on their low-carb journey.

NINA TEICHOLZ

Author of *The Big Fat Surprise* and adjunct professor, New York University

BioDiet is at once a science-based, yet easily understandable and empowering guide to learn how you can take control of your health destiny. Dr. David G. Harper and Dale Drewery have the rare gift of being able to write on a level that both engages and informs the public and healthcare professionals alike.

JIM ABRAHAMS

Founder, Charlie Foundation for Ketogenic Therapies

In *BioDiet*, Dr. Harper does a great job explaining how to reverse disease and lose weight, using food as medicine. He also shows how to make this a lifestyle with a practical plan that is easy to follow.

MARIA EMMERICH

International bestselling author of *Keto*

Dr. David G. Harper has combined what he's learned from his personal journey to improved health through dieting with his training in human physiology to produce a book that informs and guides anyone who wants to benefit from a similar experience. He explains, in simple terms, the complex biological mechanisms that are involved when we succumb to chronic disease, and how diet change can have a profound healing effect. Over the past decade, there has been an explosive growth in scientific research on the benefits of a ketogenic diet. *BioDiet* is the culmination of that knowledge in a format that is accessible and user-friendly, and will be of significant benefit to anyone who follows the diet plan.

JAY WORTMAN, MD

BioDiet

To Helen

Health & Happiness!

Health & Happiness!

+ Dale

BioDiet

The Scientifically Proven,
Ketogenic Way to Lose Weight
and Improve Your Health

David G. Harper, PhD

with Dale Drewery

PAGE TWO
BOOKS

Cataloguing in publication information is available from Library and Archives Canada.

ISBN 978-1-989025-10-9 (paperback)
ISBN 978-1-989025-77-2 (ebook)

Page Two
www.pagetwo.com

Cover design by Peter Cocking
Cover photo by malerapaso/iStock
Interior design by Setareh Ashrafologhalai
Interior illustrations by Michelle Clement

Distributed in Canada by Raincoast Books
Distributed in the US and internationally
by Publishers Group West, a division of Ingram
Printed and bound in Canada by Friesens

19 20 21 22 23   5 4 3 2 1

NOTE TO THE READER: This book is intended to educate and is not medical advice. The nutritional regimen, though scientifically validated, is not a substitute for professional medical care. It is health information. Before embarking on any significant change in nutrition, first consult with your physician, especially if you suffer from a serious ailment such as heart or kidney disease or if you are taking any medication.

I have not received any financial support from any source for the production of this book.

BioDiet™ is a trademark of David G. Harper and Pacific BioVentures, Inc.

BioDiet.org

This book is dedicated to my parents, Ruth and Charles (Mirv) Harper. My mother taught me about courage and perseverance, and my father showed me what it means to live a moral life of purpose. They are both gone now, but I think about them—and the important life lessons I learned from them—every day.

Contents

A Note from the Author
Time for a Change

'M SO GLAD you picked up this book. The fact that you spent some of your hard-earned cash to buy it and are about to invest valuable time reading it tells me that you're ready to make a change. Maybe you're frustrated: you've been eating a healthy diet and exercising regularly, and yet the scale refuses to move or, worse, moves in the wrong direction. Maybe you're feeling a bit off—tired and sluggish and lacking energy. Maybe you're worried, having been told by your physician that your blood pressure, cholesterol, or blood sugar is starting to creep up. Or maybe you're downright scared about the prospects of a life spent counting pills to control your various ailments.

As concerning as all of this might be, I encourage you to be grateful for those feelings of worry and fear. They are warning signs, and because you paid attention, they led you here. And here is where everything changes—for the better.

The BioDiet will improve your life in so many ways. It will not only help you lose weight but also reduce pain and inflammation, increase your stamina, improve your mood, and, perhaps most importantly, reduce your risk of chronic disease by as much as 70 percent. It's not an exaggeration to say that the BioDiet will lead to a whole new you.

You might be thinking, "How can a diet do all of that?"

I'm going to stop you there, because the BioDiet isn't a *diet*—at least not in the common sense of the word. It's not a quick fix to drop 10 pounds before your high school reunion or to look good at the beach. Those low-fat, calorie-restricted diets that most of us turn to when we're desperate to lose weight and inches may help our egos in the short term, but they aren't sustainable and ultimately do more harm than good. The BioDiet is a scientifically proven method to help you lose weight and keep it off while optimizing health *for your entire life*.

The "diet" in BioDiet refers to the true meaning of the word: what we eat. Specifically, the BioDiet is a low-carbohydrate, high-fat, well-formulated ketogenic nutritional regimen—optimized for our human biology—that maximizes nutrient density and incorporates targeted supplementation and super-hydration, all while taking into account individual variations and circadian rhythms.

Not exactly an elevator pitch, is it? So, let's break it down. The BioDiet is about knowing what to eat in order to give your body the fuel it needs to function as efficiently as possible. It's also about knowing *why* those food choices make sense. The more you know about how your body works, the more empowered you'll be to make it healthy and keep it that way. After all, you can't repair an engine if you have no idea what makes it run.

I'm sure you've heard about ketogenic diets; maybe you've even tried one yourself. The rates of obesity in the United States and elsewhere are skyrocketing, and people are looking for help. Lots of people, including famous athletes and actors, have tried ketogenic diets and benefited from them, so it's no surprise that the publishing

---

The BioDiet is a scientifically proven method to help you lose weight and keep it off while optimizing health *for your entire life.*

---

industry has jumped on the keto bandwagon by producing scores of books, most of them of the recipe or "how-to" variety. But the truth is that few people know how a ketogenic diet actually works. That's where *BioDiet* comes in: it will educate and empower you by explaining in a simple and straightforward way *why* ketogenic diets work and *how* to go about safely and successfully adopting one. I provide step-by-step advice from my own journey adopting and maintaining a ketogenic diet and my professional experience counseling others through the process. I also include additional nutritional advice that goes well beyond your typical diet. There's a wealth of science behind the BioDiet, but I'm fully aware that not everyone is as interested in the minutiae of cellular biology as I am. If you are, check out BioDiet.org—you'll find all the scientific details there.

By the time you're finished reading this book, you're going to know *a lot* about what makes your body's engine run and how best to support it. You're going to know what to eat and what to avoid. You're going to be able to think critically about the conflicting nutritional information out there and be empowered to make informed decisions that will affect your long-term health and well-being. And if you choose to adopt the BioDiet, you'll also be a lot leaner than you've likely been for a long time.

You picked up this book because you wanted a change, right? Let's get started.

DR. DAVID G. HARPER

Introduction
When Everything You Think
You Know Is Wrong

CAN TELL YOU exactly where I was when I realized that everything I thought I knew about healthy eating was wrong. For three years, I was the co-host of a weekly radio show called *Think for Yourself*, a program about critical thinking and reasoned skepticism. My co-hosts and I did our best to sort through the muck and the hype to find the truth underlying the issues of the day.

Several years in, we decided to do an episode on weight loss—specifically, whether it was better to diet or exercise in order to lose weight and keep it off. Our expert that day was Dr. Richard Mathias, a professor of public health from the University of British Columbia, my alma mater. We'd decided in advance that he was going to argue for nutritional intervention, while I would focus on increasing energy output through exercise. (In case you're curious, the latest research indicates that weight loss—and weight gain, for that matter—is about 80 percent diet and only about 20 percent exercise.)

In the course of our on-air discussion, I mentioned that I was familiar with the many current theories about the causes of obesity and was aware that it was a complex, multifactorial condition that resulted from interactions between physiological, psychological, social, and genetic factors. I went on for some time spouting this "party line," which still exists in most textbooks and government health sources. Dr. Mathias listened, patiently and politely. When I finally stopped talking, he spoke up. "Dave, it's not complex at all," he said. "Obesity is simply a physiological response to excess carbohydrate in the diet."

For a moment, there was radio dead air; it seemed much too simple an explanation for what we have been led to believe was a very complicated problem. It didn't take long, though, for the light bulb to go on. These days, I look back on that exchange as my "eureka" moment. All of my 30-plus years of studying human biology, health, evolution, anatomy, physiology, and pathology suddenly coalesced around Dr. Mathias's simple yet elegant statement. And in that moment, I knew he was right—absolutely right—and that I had been wrong for decades.

In my defense, I wasn't alone. For the past 40 years, the US Food and Drug Administration, Health Canada, and countless other seemingly reputable sources have recommended a low-fat, high-carbohydrate diet. The first US dietary guidelines, introduced in 1980, were designed to combat heart disease, which was then thought to be linked to a diet high in saturated fats: the so-called Diet-Lipid-Heart Hypothesis. People were advised to replace high-fat foods with those high in carbohydrates, and the processed food industry obliged by creating a host of products low in fat but high in sugar and starch. And we bought in. Today, about half of the calories in a typical American diet come from ultra-processed, high-carbohydrate foods.

The truth, however, is that those guidelines, though well intentioned, were wildly premature, released before there was conclusive

We can consider the past 40 years of nutritional recommendations as a grand experiment conducted by health bureaucrats on hundreds of millions of North Americans, investigating the effects of a low-fat, high-carbohydrate diet.

evidence to support them. In a sense, we can consider the past 40 years of nutritional recommendations as a grand experiment conducted by health bureaucrats on hundreds of millions of North Americans, investigating the effects of a low-fat, high-carbohydrate diet. And the results are in: the Organisation for Economic Co-operation and Development and US government sources report that 40 percent of American adults are now obese, a number that is expected to exceed 45 percent in the next decade. Compare this to 1970, when only about 14 percent of American adults were obese.

Obesity also has a trickle-down effect, from mother to child, and recent studies have shown that this actually begins during pregnancy. Fetuses have to adapt to the fuel supplied by the mother, and any imbalance, excess sugar for example, can cause permanent changes in fetal physiology and metabolism. For this reason, and also because children tend to eat what their parents do, they are more likely to be obese themselves. Obesity among young people, ages 2 to 19, increased from about 14 percent to nearly 19 percent in the past 20 years, and type 2 diabetes among those between the ages of 10 and 19 increased by a whopping 21 percent between 2001 and 2009. Children with obesity are also at a higher risk of asthma, sleep apnea, bone and joint problems, and heart disease. Shockingly, it is estimated that there are now 10 million children suffering from obesity-related, non-alcoholic fatty liver disease in the US alone. And, in the long term, an obese child is much more likely to become an obese adult. Sadly, these numbers show no sign of improving.

It's estimated that about 70 percent of American adults are now overweight or obese, and most suffer from insulin resistance and chronic inflammation as a result. Recent research, conducted over the past 15 to 20 years, has shown us that these conditions are all linked to a high-carbohydrate diet.

We now know that:

- cardiovascular disease is caused not by fat in our diet but by inflammation in our blood vessels;
- type 2 diabetes is aggravated by both obesity and inflammation;
- Alzheimer's disease, an inflammatory disease of the brain, is sometimes called type 3 diabetes; and
- inflammation and insulin also promote tumor growth, so a high-carbohydrate diet creates an optimal environment for cancer progression.

Obesity is not only one of the primary causes of preventable chronic disease, but it is also one of the most expensive. According to the *Journal of Health Economics*, current estimates for obesity-related healthcare costs in the United States range from $147 to $210 billion per year. Job absenteeism as a result of the disease adds another $4.3 billion annually. In Canada, the direct and indirect costs of obesity on the already overstretched healthcare system are expected to reach $33 billion by 2025.

Kind of hard to believe we ever thought a diet rich in carbohydrates was a good idea, isn't it?

Fortunately, there is good news. Greatly restricting dietary carbohydrate through the adoption of a ketogenic diet will reduce mid-abdominal fat and the inflammation and insulin resistance that goes along with it. The result is the significant reduction, and sometimes even the elimination, of chronic disease by:

- reducing inflammation in the blood vessels, which causes heart disease;
- lowering and stabilizing the high blood-glucose levels responsible for diabetes;

> If you're willing to commit
> 12 weeks to changing
> the way you eat, you really
> will change the rest
> of your life for the better.

- reducing inflammation in the brain that contributes to Alzheimer's disease; and
- reducing the glucose and insulin that promote the growth of cancer cells.

And a bonus bit of good news? A ketogenic diet not only helps prevent these chronic diseases, but it can also help treat those already afflicted.

I learned all of this during a period of intense research following my "eureka" moment with Dr. Mathias. I wanted to be sure of the truth. I wanted to know if his simple explanation for our expanding waistlines—and the health consequences that go along with them—would hold up to serious scientific scrutiny. And so I got busy. I put aside everything I thought I knew about healthy eating to investigate the benefits of low-carbohydrate, high-fat diets in preventing and treating chronic disease. The BioDiet is the result of that research— the book, the way of eating, and the lifestyle it inspired.

As a scientist and health educator—an associate professor of kinesiology at the University of the Fraser Valley, a visiting scientist at the BC Cancer Research Centre, a scientific advisory board member of the Canadian Clinicians for Therapeutic Nutrition, and a member of the Institute for Personalized Therapeutic Nutrition— I want to set the record straight. I want to share the recent, robust, and conclusive scientific research that explains why our traditional way of eating has been failing us, and why a high-fat, low-carbohydrate diet is a much healthier choice. I adopted a ketogenic diet in 2012, initially as an experiment to see if it actually achieved the results my research indicated it would. The experiment soon became a way of life, and today I continue to reap the benefits of lower body weight and better overall health.

In the pages that follow, I introduce you to the BioDiet so that you, too, can enjoy its benefits. I debunk some of the popular food myths that have gotten us into this overweight, inflamed, and insulin-resistant mess. I explain how and why the BioDiet combats obesity, helps reverse some of the effects of aging, and reduces the incidence and severity of chronic disease. And, perhaps most importantly, I give you the tools you need to determine if the Bio-Diet is right for you. If it is right for you (and for most of us, it *really, really* is), I go over a plan to incorporate it, step by step, into your life. I lay out the five steps—Bio-Assessment, Bio-Preparation, Bio-Adaptation, Bio-Rejuvenation, and Bio-Continuation—that will help you to heal your body and embark on a healthier future. And I set you up with resources, including menu plans, shopping lists, and supplement suggestions, to make the transition to this way of eating as easy as possible.

When I speak publicly on the BioDiet, I say, "Give me an hour and I'll change your mind. Give me 12 weeks and I'll change your life." So, what do you say? Do you want to reduce your weight; lower your pain from systemic inflammation; stabilize your blood sugar; improve your immune function, your energy, and your mood; and slow your aging process? If you're willing to commit 12 weeks to changing the way you eat, you really will change the rest of your life for the better. Not a bad deal, right?

1

Why— What the Science Says

Out of Our Element
Why We Need the BioDiet

Science progresses one funeral at a time.
MAX PLANCK, GERMAN PHYSICIST, 1858–1947

PAINTED A PRETTY bleak picture in the introduction. Rising obesity rates. Rising chronic disease rates. Rising healthcare costs. It's a scary time we're in, especially when you consider how much attention is paid in the media to living our best and healthiest lives. Magazines, books, television shows, blogs, social media, YouTube channels—you'll find more health advice than you can possibly take in. So, where's the disconnect? How can we possibly be so health-obsessed and yet so out of touch with what our bodies actually need to be healthy? The answer is remarkably straightforward: we're out of our element.

Once upon a Paleolithic Time

In our ancestral, hunter-gatherer days, there was no breakfast cereal, no pasta, no panini. Although we don't know precisely what our ancestors ate, we can speculate based on our knowledge of what plants and animals were around at the time, the tools and cooking implements they left behind, and the oral histories of present-day Indigenous peoples. It's safe to say that our diet was opportunistic, mostly derived from whatever animals we could catch and whatever vegetation we could stomach. What little sugar we got came from berries and other sparsely available seasonal fruits, as well as the occasional bit of honey, if we were prepared to brave the bee stings. Food scarcity was a real issue, but we learned to cook meat over fire and to dry fish and fruit in the sun, preserving them for times when food was not plentiful. Our diet was largely animal-based, high in fats, and very low in carbohydrates. We ate this way for more than a dozen millennia: it is our natural state as humans.

In our ancestral, hunter-gatherer days, there was no breakfast cereal, no pasta, no panini.

If you think this is all ancient history, think again. There are still a few remaining hunter-gatherer peoples from whom we can learn. On their traditional diets, they generally live long, disease-free lives, provided they can harvest enough to eat from their environment and avoid hurting themselves in the process. For example, Inuit of Northern Canada on their traditional diet, which can be upwards of 70 percent fat, rarely suffer from obesity, diabetes, cardiovascular disease, or cancer. Most of their calories come from marine mammals and fatty fish, and they eat almost the entire animal, savoring as delicacies the nutrient-rich organs. The traditional diet of East Africa's Maasai people consists almost exclusively of raw meat, milk, and cow's blood. Like Inuit, Maasai are largely free of chronic disease—until they eat a Western diet.

Dawn of the Bread

So, what happened? How did so many of us go from eating a healthy diet to one that causes chronic disease? Well, it started about 10,000 years ago thanks to a development made with the best of intentions: the onset of agriculture. The cultivation of plants, particularly cereal grains, was a hedge against famine. Like dried or salted meat and fish, small quantities of grains could be transported, and many were high in carbohydrates, which provided ready calories during times of scarcity.

Over time, improved agricultural techniques like irrigation yielded surpluses of food that required storage, encouraging humans to stay put. We learned to grind grains to make breads (and we learned to make beer, too!). Communities grew and wealth increased. Overall health, however, did not. Ethnological and archeological studies reveal that the transition to cereal-based diets resulted not only in the decline of people's health but also their height. By significantly reducing the diversity of our diet, this stable, carbohydrate-rich food source had increased nutritional deficiencies, making us more susceptible to stress and more likely to die at a younger age.

Diseases caused by nutritional deficiencies, including those that are the result of a carbohydrate-rich diet, can take decades to fully develop; you have to live long enough to see the effect. In Europe, from 1750 to 1850, life expectancy hovered between 40 and 45 years of age. Most people died young from infectious diseases, accidents, and wars. Chronic diseases were the unfortunate fate of the fortunate few who lived into true old age. By the end of World War II, life expectancy in the United States was approximately 60. In recent decades, that has risen dramatically, mostly due to improvements in medical science, which resulted in a decrease in death from acute and infectious diseases.

High-Carb Planet

Today, just about every culture on the planet has at least one starch as a dietary staple. Corn, which was domesticated in present-day Mexico, has spread throughout the world thanks to international trade and now comprises a major food source in Africa, Europe, and the rest of North America. Rice is the main source of nourishment for more than 1.6 billion people around the globe, from Asia to Latin America to Africa. The largest producers of wheat, which originated in present-day Turkey, now include the European Union, China, India, Russia, and the United States. According to the Food and Agricultural Organization, corn, rice, and wheat together comprise 51 percent of the world's calorie consumption. Root and tuber crops, including potatoes and cassava (which when dried and ground is called tapioca), make up 5.3 percent of the world's nourishment.

Today's baby boomers, now in their 60s and 70s, will likely live into their 80s and 90s—certainly old enough to experience the chronic diseases that are the result of decades of eating high-carbohydrate foods. Here's the good news, though: by significantly reducing carbohydrate consumption, they can ameliorate and even reverse chronic illness—a benefit not just to the boomers but also to our healthcare system as a whole.

Back to the Future

Low-carbohydrate diets are by no means new in the Western world; they've been around for more than 150 years. The first on record was prescribed in 1862 by British surgeon William Harvey for his obese patient William Banting, an undertaker to the wealthy (evidently, he made the coffin for the Duke of Wellington). Like many of us, Banting had begun slowly gaining weight in his 30s. By age 62, he weighed 202 pounds.

Not so bad, you're thinking? However, Banting was just five-foot-five and having trouble with, well, normal daily routines:

> *I could not stoop to tie my shoes, so to speak, nor to attend to the little offices humanity requires without considerable pain and difficulty which only the corpulent can understand. I have been compelled to go downstairs slowly backward to save the jar of increased weight on the knee and ankle joints and have been obliged to puff and blow over every slight exertion, particularly that of going upstairs.*

Dr. Harvey's prescription for weight loss, which he gave to Banting in August 1862, was simple: avoid foods of a "sugary or starchy nature." By Christmas of that year Banting had shed 18 pounds, and by the following August he was down to 156 pounds and had lost

12 inches of belly fat. He had regained the ability to go about his life and "attend to the little offices" that had become increasingly difficult over the 30 years of weight gain. And he was delighted to have achieved all of this on a diet he'd once considered indulgent and unwise:

> *I can conscientiously assert I never lived so well as under the new plan of dietary, which I should have formerly thought a dangerous, extravagant trespass upon health . . . I am very much better both bodily and mentally and pleased to believe that I hold the reins of health and comfort in my own hands.*

Banting was so thrilled about his return to good health that he wanted to share his newfound wisdom with others. In 1863, he wrote a short booklet called *Letter on Corpulence, Addressed to the Public*. He had 1,000 copies printed and gave them away at no cost. Billed as the world's first diet book, it has been reprinted many times, most recently in 2007. In it, Banting describes his personal experience with the diet and provides details of what he ate at each meal. His diet was decidedly low-carbohydrate. He eliminated pastry, milk, beer, and champagne (la-di-da), most fruit, almost all bread, and potatoes, although he did eat other vegetables, including the Brits' cherished carrots and peas. The diet came to be known as a *bant*, and to partake was called *banting*.

The bant made its way around the world and, for about a century, was commonly prescribed to the obese. The term *bant* is still in use in Sweden, where *banta* means to "slim" or to "cut back." William Banting continued on his diet for the remainder of his life. He died in 1878, at the ripe old age of 81.

Nearly three decades later, a 24-year-old graduate student embarked on a low-carbohydrate regimen but for entirely different

reasons. In 1906, Harvard-educated anthropologist Vilhjalmur Stefansson joined a polar expedition to live with Inuit for a summer. But
when he missed the planned rendezvous to return home, Stefansson was forced to winter in the north. For months, he ate only what
Inuit ate, a diet composed entirely of meat and fatty fish. His south-
ern colleagues were seriously concerned that Stefansson would die
from malnutrition due to a lack of carbohydrate, but when he finally
did make it home, the young explorer was as healthy as ever and
considerably leaner. Some years later, Stefansson decided to return
to the north and repeat the experience, this time with colleague Dr.
Karsten Anderson. The two stayed for a full year and, once again,
ate only the low-carbohydrate, high-fat Inuit diet. And, as before,
Stefansson, and now Anderson, returned home in excellent shape.

The pair's colleagues remained skeptical about the benefits of
such a diet, so in 1928, Stefansson and Anderson conducted a super-
vised experiment in the hopes of convincing them, once and for all.
For one year, while under medical oversight at New York City's Bel-
levue Hospital, the two ate an Americanized version of the Inuit
diet, which included few vegetables or fruit. During that time, the
pair devoured steaks, chops, poultry, organ meats, fish, and fat. The
results, published in 1930 in the *Journal of Biological Chemistry*, were
conclusive: not only were both men perfectly healthy at the end of
the 12-month period, but the low-carbohydrate, high-fat diet had, in
fact, *improved* their health. Stefansson and Anderson's experiment
revealed that we can get all of the vitamins and minerals we need
from animals, so long as we eat the whole animal as Inuit do. It is
unknown whether the pair continued to adhere to a low-carb diet,
but Stefansson did write a food book in 1946 called *Not by Bread
Alone*. He died of a stroke in 1962 at the age of 82.

Top 10 Food Myths

There are enough food and nutrition myths to fill an entire book of their own. To save us a bit of time, here are the top 10, along with the real story based on the latest science.

Myth #1: Eating fat will make you fat.

Truth: It's true that fat is denser in calories than carbohydrates and proteins (more than twice the calories per gram), but obesity is not primarily due to an excess of calories consumed. It is the *type* of calories consumed that is important. Recent science shows that most surplus weight and obesity is caused by excess carbohydrates in the diet. Fats are a main source of energy. They are also an important source of fat-soluble vitamins and provide much of the pleasurable flavor and texture in food. Some (the omega fats) are, in fact, essential in our diet, as we can't produce our own.

Myth #2: Saturated fats are bad for your heart.

Truth: There has never been any robust, conclusive evidence that saturated fats cause chronic disease. In fact, saturated fats are the cleanest-burning fuel you can put in your body. From a health perspective, saturated fats are not only benign, they're beneficial.

Myth #3: Carbohydrates are essential to our bodies.

Truth: There are no essential carbohydrates. Your body evolved to make its own blood glucose from non-carbohydrate sources. When it does so, it makes the optimum amount for the present needs of the body. There are beneficial carbohydrates—soluble and insoluble fibers—but you can get plenty of these without also burdening your body with sugars and starch.

Myth #4: Gluten-free eating is the healthiest option.

Truth: If you have celiac disease or are gluten sensitive, by all means avoid gluten in your diet. Otherwise, keep in mind that most processed, gluten-free foods use substitutes like rice flour, potato starch, and tapioca flour. These and other starches rapidly raise blood glucose and insulin, aggravating diabetes and other chronic diseases. Gluten-free does not mean low-carbohydrate. In fact, it's sometimes quite the opposite.

Myth #5: Everything in moderation.

Truth: To quote Canadian physician Dr. Jay Wortman, "Everything in moderation is an excuse we use to eat the things we shouldn't eat." Like the notion of a "balanced diet," "everything in moderation" gives us license to trade off nutritious calories for empty ones. This is doubly dangerous when that junk food contains sugar, which activates the opiate receptors in our brain, stimulating our reward center. Each time we eat something sweet, we're reinforcing those neuropathways and hardwiring our brains to crave the stuff. So, next time you catch yourself using "moderation" and "balance" as a rationale to consume foods you know are bad for you, it helps to remember that you're not only fooling yourself, you're compromising your health in the process.

Myth #6: To lose weight, you need to cut calories.

Truth: Cutting calories means you eat less food, and eating less food means you have less of an opportunity to meet daily nutritional requirements. If you are restricting calories to less than your daily needs, you will not only be perpetually hungry, but you will also reduce your metabolic rate, making weight loss more difficult. What's more, once you return to your regular diet, there is a high probability that you will regain the weight you lost and are likely to put on even more. See "Yo-Yo Dieting" in chapter 9 (page 193).

Myth #7: Fruit is good for you because it's natural.

Truth: Newsflash: fruit did not evolve to be a health food. Its evolutionary imperative is to spread its seeds, and the best way to do that is to get animals to eat it, move on, and deposit the seeds, some distance away, embedded in a healthy dollop of fertilizer. Sweet fruit is more attractive to animals—including humans—so job well done on the dispersal-system front. But the sweetness comes at a high cost not only in terms of high-carbohydrate starches but also fructose—a known toxin. The same goes for honey and maple syrup. Don't be persuaded to buy and eat food simply because it's considered natural.

Myth #8: All vegetables are created equal.

Truth: Many vegetables—especially root vegetables, beans, and grains (and, yes, I include grains as vegetables because they are plants)—are high in starch and can contribute to obesity and insulin resistance. Choose wisely.

Myth #9: Ketogenic diets don't provide sufficient nutrients and fiber.

Truth: The BioDiet is specifically designed to ensure that you get more than enough essential proteins, fats, and vitamins. And, because it's rich in plant-based foods like nuts, seeds, vegetables, and berries, the BioDiet also provides plenty of fiber, especially soluble fiber, which has been identified as having a variety of health benefits.

Myth #10: If you work out, you can eat whatever you want.

Truth: Working out does burn calories, so your food intake should increase proportionally. However, science tells us that about 80 percent of weight management is determined by what you eat, not how many calories you burn. You lose weight in the kitchen; you get fit in the gym. If you eat poorly, exercise will not help you outrun the negative health consequences.

Off Track

Stefansson and Anderson's experiment should have been enough to give us pause and perhaps even return us to a way of eating more in tune with our biology. Unfortunately, it wasn't to be. This prescription for weight loss and good health was lost to us in 1955, when Ancel Keys of the University of Minnesota introduced a new theory about the cause of heart disease—the so-called Diet-Lipid-Heart Hypothesis. Based on his research in seven countries, Keys concluded that cholesterol from foods high in saturated fat was the cause of cardiovascular disease. He even went so far as to develop an equation that predicted blood serum cholesterol based on the consumption of saturated versus unsaturated fats. The Diet-Lipid-Heart Hypothesis, though never validated by the scientific community, was nonetheless widely accepted—and it gained additional momentum following US president Dwight Eisenhower's heart attack in September 1955. According to Keys, the president's heart problems were the result of his high-fat diet, not his several-packs-a-day smoking habit. (Keys can be forgiven for this error; at the time, we didn't know that cigarette smoking caused heart disease.)

For years, Keys, along with his like-minded colleagues, dominated the discussion about the causes of heart disease. Critics of Keys's work had their careers derailed and found it increasingly difficult to access government funding for research that could scientifically debunk his findings. Even today, despite a clear lack of supporting evidence, the demonization of fats, especially saturated fats, persists, and government agencies continue to financially favor research that vilifies fat and celebrates carbohydrates. Little wonder: to do otherwise would be colossally embarrassing. Not only would governments need to acknowledge that they were profoundly wrong about the supposed benefits of a high-carbohydrate diet, they'd also have to admit to being in the pocket of the powerful processed-food

Even today, despite
a clear lack of supporting
evidence, the demonization
of fats, especially saturated
fats, persists, and government
agencies continue to
financially favor research
that vilifies fat and celebrates
carbohydrates.

Vested Interests

Nutrition may be based on science, but make no mistake: it's also influenced by business, politics, and personal ambition. Those with vested interests play an enormous role when it comes to the food we put in our mouths every day. For example, a nutrition scientist who has spent decades researching from a particular perspective—say, that fats are bad for you—has a lot riding on that point of view, including their career and their livelihood. They have reason to stand their ground, even when it shakes beneath their feet.

The processed food industry is perhaps the group with the largest vested interest in maintaining the status quo. This is why grocery stores are full of highly profitable junk food designed in taste and texture to make us eat more. Seventy-four percent of packaged food sold in supermarkets contains sugar, including salad dressing, pasta sauce, and ketchup. Interspersed with these sugar-laced products are equally profitable fat-free ones that the industry promises will help us lose the weight we gained eating the junk food! It's duplicity at its most delicious.

If you think our governments will help protect us from the self-serving food companies, you might be disappointed. The food industry is an enormous contributor to the economy, which gives companies the clout to influence important initiatives like government food guides and dietary recommendations. And these are the nice people in this story. Thanks to the internet, we are bombarded by messages from those looking to make a quick buck on the latest "health" food, "energy" drink, or "nutritional" supplement. Largely unregulated, this realm is filled with many charlatans who misrepresent science and fabricate the benefits of their product. In short, these people prey on our hopes, convince us of their lies, and pry open our wallets in the process.

industry, which has poured untold billions into low-fat food production, often replacing healthy fats with high-carbohydrate fillers. And so, unfortunately, many of the dietary myths that grew out of Keys's untested hypothesis stubbornly persist today (see "Top 10 Food Myths," page 18).

Although Ancel Keys's Diet-Lipid-Heart-Hypothesis continues to have its supporters, there are fewer of them with every retirement. His name, or at least his initial, lived on, however, in the form of the K-ration, which Keys developed for the military with the help of chewing gum industrialist William Wrigley and the Cracker Jack Company. Designed to feed soldiers on the battlefield, the portable, non-perishable K-rations contained hard biscuits, dried sausage, candy, and chocolate—a significant 3,200 calories per day, the vast majority of them carbohydrates.

A New Hope

In 1972, physician Robert Atkins published a book called the *Dr. Atkins' Diet Revolution*. In it, Atkins described his success in treating the overweight, including himself, with a low-carbohydrate nutritional regimen high in protein, including meat, cheese, and eggs. The book quickly became a bestseller, eventually selling more than 10 million copies worldwide. But critics argued that while the diet worked in the short term, and millions of Americans had lost weight, it was unsustainable and the high-protein component of the diet could lead to serious liver and kidney problems. Although in his subsequent books Atkins stressed the importance of incorporating lots of fats, vegetables, and fruits into the diet to help avoid any medical concerns, the damage to his reputation was done. Atkins and his diet had further engrained the belief among many, including government officials and the medical world, that a low-carbohydrate diet was unhealthy or, worse, downright dangerous.

Fortunately, there is a sea change underway. We are finally publicly acknowledging that it is indeed our over-consumption of carbohydrates, not fat, that's making us sick. And low-carbohydrate diets, especially ketogenic ones, are being recognized as a legitimate dietary intervention for dozens of diseases, including obesity and diabetes. For example, in September 2018, a group of more than 4,000 physicians, researchers, and nutrition practitioners appealed to the Canadian government to consider the most current scientific research available before, once again, eschewing fats in favor of carbohydrates. In their petition, they cited a July 2018 study that investigated links between dairy fat consumption and chronic disease. The study, which was published in the *American Journal of Clinical Nutrition*, followed nearly 3,000 adults for more than 20 years.

The authors of the study concluded:

> For decades, dairy fat consumption has been hypothesized to be a risk factor for cardiovascular disease, as well as potentially diabetes, weight gain, and cancer. Little empirical evidence for these effects existed from studies of clinical events. In current years, a growing number of prospective studies have shown generally neutral or protective associations between self-reported dairy foods and dairy fat consumption with the risk of cardiovascular disease, diabetes, weight gain, and cancers, raising questions about conventional wisdom.

Today, in just about every Western nation, there are large and quickly expanding networks of health professionals—physicians, dieticians, nutritional counselors, and researchers—working hard to convince their governments to recommend diets that are supported by rigorous, compelling scientific research.

So, how do you separate fact from fiction, and decide for yourself what's best for you? That's where this book comes in. *BioDiet* advocates, in part, a return to our pre-carbohydrate, hunter-gatherer days, but unlike diets based largely on anecdotal evidence, the BioDiet incorporates what the latest, most robust science tells us about optimal nutrition. The result is a well-formulated ketogenic diet that provides complete nutrition without restricting calories. It embraces healthy fats and moderate protein consumption, significantly reduces carbohydrates and their negative impact on our health, and results in the production of beneficial ketones (more on those in upcoming chapters). Table 1.1 compares the BioDiet to some of the other low-carbohydrate diets out there.

Table 1.1: A comparison of low-carbohydrate diets. It is important to note that while all ketogenic diets are low-carbohydrate, not all low-carbohydrate diets are ketogenic, and not all ketogenic diets are the same.

	Whole30	Paleo Diet	Atkins Diet	BioDiet
Macronutrients				
Fat %	50	65	60	70
Protein %	30	15	35	25
Carbohydrate %	20	20	5	5
Ketogenic?	**No**	**No**	**Maybe**	**Yes**
Other characteristics				
Fruit	Yes	Yes	No	Limited to berries
Dairy	No	No	Yes	Yes
Sugar	Yes	Yes	No	No

I've spent years diligently researching this new science of nutrition with an open but skeptical mind, and I have come to the conclusion that we are now entering an era in which we are finally acknowledging that what we eat is an important, if not the most important, factor in our physical and mental well-being. But I don't want you to mindlessly adopt a way of eating that I believe has the power to change your life. I want you to understand *why* it has that power, and why it's hands down the best thing you can do for yourself and your health.

Read on.

We are now entering an era in which we are finally acknowledging that what we eat is an important, if not the most important, factor in our physical and mental well-being.

2

It's All about the Fuel
Keeping Your Motor Running

The flame that burns twice as bright burns half as long.
LAO TZU, CHINESE PHILOSOPHER,
SIXTH CENTURY BCE

ROM THE MOMENT we come kicking and screaming into this world, we know what it is to be hungry. Newborns cry out to be fed, baby birds are starving from the moment they hatch, and a newborn kangaroo—blind, hairless, and only a few inches long—is so hungry it's prepared to climb through the thick fur on its mother's abdomen to reach her pouch to feed. As animals, we eat in order to survive. Our bodies take the food, metabolize it, and use the energy and raw materials to grow, repair, and regenerate.

Understanding this process—and how it can change depending on what type of fuel we put into our bodies—is integral not only to knowing how the BioDiet works but also to making important decisions about what we eat. And that process is precisely what we focus on in this chapter. We start with a quick refresher course in how our bodies build and refuel themselves. We look at what's in the food we eat and the impact of proteins, carbohydrates, and fats on how our bodies function. We also explore the importance of hormones such as insulin, cortisol, growth hormone, and the ones that determine whether we feel hungry or full. We learn about our second brain and why it's almost as important as the one that's between our ears. Finally, we explore just what happens when we fuel ourselves with the high-carbohydrate, low-fat diet to which we've grown accustomed. So, let's brush up on that biology!

Macronutrients: The Building Blocks

Macronutrients are the chemical compounds that we consume in large quantities—hence the "macro." These organic (carbon-containing) compounds are vital to our survival, because we use them for energy and as structural building blocks in our cells. There are three types of macronutrients: proteins, carbohydrates, and lipids (which include fats). To understand how these groups differ and how they interact, imagine every cell in your body as a cabin in the woods. Our lives depend on how we build that cabin, what's inside, and how we keep the home fires burning.

Proteins

Proteins are the macronutrient world's heavy lifters. They define who we are and how we function. They account for most of the materials in our "cabin" and provide the tools we need to construct it and its contents. Proteins make up the internal scaffolding of cells as well as external structures that form parts of our skeletal system and skin. They regulate the movement of material into and out of each cell, and they serve as messengers for cell-to-cell communication. They also make up the filaments within muscle cells that allow them to contract. I wasn't kidding when I said they were heavy lifters.

A typical cell contains roughly 100,000 different proteins, and we only know what roughly half of them do. What we *do* know is that all proteins are made up of building blocks called amino acids, which consist of carbon, hydrogen, oxygen, nitrogen, and, in some cases, phosphorus and/or sulfur. There are 20 different types of amino acids, but our bodies can only produce 11. We call the other nine *essential*, because it's essential that we get them from our diet. Once our bodies have broken down the proteins in our food into amino acids, they are absorbed into the blood and transported to our cells, where the building blocks are reconstructed into new proteins to provide the cells with their structures and facilitate their functions.

Carbohydrates

The primary function of carbohydrates is as an energy source for our cabins; they are the fuel we feed to the wood stove. In our cells, they also form some internal structures and are responsible for some important functions. All of our organs need some carbohydrates to function properly, especially the brain. Red blood cells need them, too. Like proteins, carbohydrates also have basic building blocks: *monosaccharides*, from the Greek words *monos* ("single") and *sacchar* ("sugar"). Monosaccharides are the simplest forms of sugar, but depending upon their structure, they form glucose, fructose, or galactose. Of these, glucose is the most important. It circulates in our blood as blood sugar, supplies almost all the fuel for our brain, and is the principal source of energy for metabolism in other cells. When our glucose levels get out of whack, so do we.

Disaccharides are made up of two sugars. We know the disaccharide sucrose, made from glucose and fructose, as table sugar.

Most table sugar is extracted from sugarcane or sugar beets, but sucrose is also found in most fruits, sometimes in high concentrations. Lactose, which is glucose bonded to galactose, is milk sugar (*lac*, the Latin word for "milk," plus *ose*, which is used to name sugars). Those who are lactose intolerant are unable to break the bond that joins the glucose and galactose, which can result in some pretty uncomfortable gas production by the gut bacteria. The disaccharide maltose is made from two glucose molecules. Although it's not naturally found in large quantities in our food supply, our bodies do generate maltose when we digest grains, vegetables like sweet potatoes, and some fruits, especially peaches and pears. In blood circulation, it's a way to transport glucose.

Polysaccharides are made up of many glucose molecules, typically thousands, all bonded together. These "complex carbohydrates" include starches—which come from plants such as wheat, oats, barley, corn, and rice, to name a few—and the products we make from them, like bread and pasta. Legumes (for example, beans and peas) and root vegetables are also high in starch. Glycogen is also another important polysaccharide, similar in structure to starch but found in animals. We store glycogen in significant amounts in our muscles, where we can quickly access it and turn it back into glucose for energy.

Complex carbohydrates also include fiber, which comes in two varieties, both of which are found in varying amounts in all fruits and vegetables. Soluble fiber, which can be digested by bacteria in our large intestine and converted into beneficial fats, is found in vegetables such as broccoli and carrots; fruits, including figs, avocados, and berries; and nuts, especially almonds. Insoluble fiber, which is indigestible and makes up the bulk of our feces, is found in whole grains and crunchy vegetables like cauliflower, legumes, and seeds. Some fruits—plums and prunes, for example—have a foot in both worlds: their pulp is a soluble fiber, but their thick skin is insoluble.

Figure 2.1: Types of carbohydrates found in our diet.

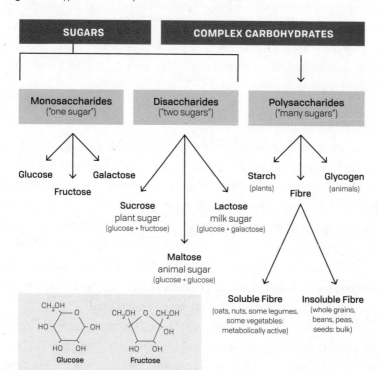

Back to our cabin—and its wood stove—for a moment. Remember each cell in our body contains its own stove, and each of those stoves needs to be fed. So, what kind of fuel should we be using to get the job done? Our bodies can easily burn carbohydrates (as glucose), but this is a bit like relying on kindling to keep a blaze going. Like kindling, carbohydrates burn bright and burn quickly, which is why you get hungry an hour after eating a carbohydrate-rich breakfast (like toast, juice, and cereal). In addition to creating a need to constantly feed the fire to keep our cells functioning, high-carbohydrate consumption results in another issue: the production of oxidants. Over time, oxidants can damage cell structures, including the DNA found in the cell's nucleus. This contributes to degenerative diseases and aging, which is one reason why many people take antioxidants as part of their daily vitamin regimen. So, what's the alternative to eating carbohydrates to keep our cellular cabins functioning properly? The answer lies with our next macronutrient—lipids.

Lipids (Including Fats)

Some incorrectly use the term *fats* for this category of macronutrient, but fats are actually a subgroup of the larger lipid category, which includes all of the water-insoluble organic compounds. We do refer to fats alongside carbohydrates and proteins as macronutrients, but lipids also include important structural compounds that make up much of the internal and external membranes of cells. Another class of lipids, steroids, are hormones derived from cholesterol, including cortisol and the sex hormones. Some important cell-signaling compounds are also classified as lipids.

For our purposes, however, we'll focus on fats, which are more than twice as energy dense as proteins and carbohydrates. Dietary fats are technically called triglycerides, because they consist of three fatty acids: long chains of carbons with hydrogens attached to a single glycerol, like fingers held up at a crowded bar to order three more beers. Adipose tissue, commonly called body fat, is specifically designed to capture and store excess energy as triglycerides.

There are two types of fatty acids that give rise to the names we commonly hear for fats: *saturated* and *unsaturated* (see figure 2.2). The difference between the two has to do with their carbon chains. Like Tinkertoys, carbon is able to form chains with up to four other molecules, including other carbons. Sometimes these bonds double up, so the carbon bonds with fewer other molecules (see inset, figure 2.2). Saturated fats, which are found in meats, dairy, and some plants, contain only single bonds and, as a result, have the maximum number of hydrogen atoms attached (hence "saturated"). Since most of a macronutrient's energy is found in the bonds between carbon and hydrogen, saturated fats are the most energy dense. They also resist oxidization, which, outside of the cells, can lead to the formation of chemicals that are harmful to our health.

Unsaturated fats contain at least one carbon double bond in their fatty acid chains. The position of the carbon double bond defines the type of unsaturated fatty acid. The fatty acid chain has an alpha end and an omega end (alpha being the first, and omega being the last, letter of the Greek alphabet). For example, referring to the unsaturated fatty acid illustrated on the inset of figure 2.2, you will see the first double bond on the third carbon counting from the omega end. That makes this an omega-3 fatty acid.

In a healthy diet, the ratio of omega-6 to omega-3 should be about 2:1, as was the case 50 years ago. That ratio now exceeds 16:1!

Figure 2.2: Types of fatty acids in our diet.

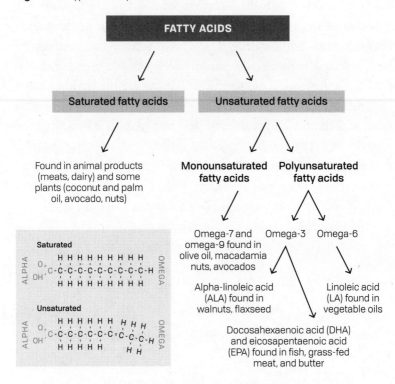

Depending upon the number of double bonds, unsaturated fats are either monounsaturated, meaning a single double bond, or polyunsaturated, meaning more than one double bond. Some polyunsaturated fats, like those found in vegetable oils, contain omega-6, which is pro-inflammatory. Omega-3, which is found in fish oil, grass-fed meat, butter, and flaxseed oil is anti-inflammatory. In a healthy diet, the ratio of omega-6 to omega-3 should be about 2:1, as was the case 50 years ago. According to the British medical journal *Open Heart*, that ratio now exceeds 16:1! No wonder we're chronically inflamed.

To assess the levels of lipids in the blood, we measure triglycerides and cholesterol. Low-density lipoproteins (LDLs) are often referred to as "bad cholesterol," whereas high-density lipo-proteins (HDLs) are seen as "good cholesterol." Physicians often use these clinical indicators, referred to as biomarkers, to determine a person's risk of cardiovascular diseases like hypertension, heart attack, and stroke. However, the predictive value of these numbers is now being questioned.

Nucleic Acids

Though not considered macronutrients, nucleic acids—a fourth category of organic compounds—make up the genetic material in our cells. They consist of carbon, hydrogen, oxygen, nitrogen, and phosphorus and are comprised of basic building blocks called nucleotides. These assemble into deoxyribonucleic acid (DNA) and ribonucleic acid (RNA) in our cells. The simplest way to describe the function of nucleic acids is to think of DNA as a recipe book for making proteins and RNA as the recipe, chef, and kitchen. Nucleic acids are found in all cells, so we get plenty of them in our food. However, we don't consider them in a nutritional sense, only a functional one.

Hormones: Not Just about Sex

Ask a high-school biology class about hormones and chances are you'll be met with some snickering and kids staring at their shoes. But as we know, hormones aren't just about sex, acne, and mood swings. They are essential to all manner of important bodily functions, from heart rate to digestion to metabolism. Produced by organs called glands, hormones are secreted into our blood and dispatched to cells elsewhere in the body. Some, like sex hormones, are targeted to specific tissues, while others, like growth hormone, have the capacity to affect all tissues. Let's start our exploration with insulin, the principal hormone affecting how our body fuels itself.

Insulin

Discovered in the 1920s by Sir Frederick Banting (no relation to our undertaker, William Banting, and his bant legacy), insulin has four important functions, foremost of which is helping our cells absorb glucose. This is crucial for our glucose-dependent brain cells, which is why the first signs of low blood glucose are confusion and dizziness. Deprived of glucose, brain cells will die. Same goes for red blood cells. Insulin further reduces excess glucose in the blood by promoting glucose conversion to triglycerides, which are stored as fat in our adipose tissue. Second, along with growth hormone, insulin is important for the production of proteins, which, as discussed earlier, are the main structural and functional components of cells. Third, insulin affects our mood, memory, and other cognitive functions, including hunger and satiety. Finally, an oft-forgotten but important function of insulin is sodium regulation. It causes the kidneys to retain sodium, and when they do, they reabsorb water as well.

Cortisol

Like all steroid hormones, cortisol is derived from cholesterol. We use a variant of this hormone (hydrocortisone) as a medicine. It is a powerful anti-inflammatory that can be injected into painful joints or used topically to treat many skin ailments. Cortisol is also a stress hormone that is released in severe circumstances: physical or psychological stress, starvation, dehydration, sleep deprivation, extreme temperatures, or following an injury. In situations like these, cortisol mobilizes the building blocks of our organic compounds and dumps them into the blood: fat that's stored in the adipose tissue is released as triglycerides and fatty acids, glycogen is converted into glucose, and proteins are broken down into amino acids. This makes the blood into a rich soup of energetic molecules and building supplies readily available to go where needed and address the stressor or repair damaged tissues.

Growth Hormone

As its name suggests, growth hormone is responsible for stimulating and regulating the growth of our tissues. Although we require higher amounts of growth hormone when we are young, it continues to

What Color Is Your Fat?

There are two types of adipose tissue—white fat and brown fat. White fat is what most of us have too much of. Some of it is subcutaneous, which means it lies just under our skin. There doesn't seem to be any negative consequence to subcutaneous fat in reasonable amounts. In fact, it acts as an insulator, provides a cushioning layer to protect underlying tissues, and helps with hormone metabolism. However, white fat that collects in large amounts on the torso—especially visceral fat that forms in our mid-section, around the internal organs—is another story. Fat deposited here is the unhealthiest "normal" tissue you can have. Fifty percent of the cells here are actually specialized white blood cells called macrophages and they are a major source of systemic inflammation (see chapter 3 for more on inflammation).

Until fairly recently, we had no idea that brown fat existed. It's the type of fat found on babies, especially near the back of their necks. Brown fat is very metabolically active and, as a result, produces a lot of heat. Think of it as nature's hot water bottle, keeping little ones warm. Most of us lose our brown fat as we mature, but some lucky skunks maintain a fair amount. Since brown fat helps burn calories, these folks can eat pretty much whatever they want and never gain weight. I hate them.

be released throughout our lives. The main function of growth hormone in adults is to stimulate protein synthesis, so we can build new cells in our blood, skin, and digestive system; repair damaged muscle fiber; help keep our bones and other connective tissues strong; make intracellular and extracellular structures; and create molecules for cell-to-cell communication, among other things. Without sufficient growth hormone, we fall apart faster than we can build ourselves back up, and our body doesn't function well.

Leptin and Ghrelin

Two other hormones that play important supporting roles in how we metabolize food are leptin and ghrelin. Leptin, from the Greek *leptos* meaning "thin," is produced by adipose cells in our fat and works to inhibit hunger. It is responsible for the feeling of being satiated—of having eaten enough. It seems that many obese people have become insensitive to leptin and so tend to overeat, aggravating their obesity. Ghrelin, which is produced in our digestive tract, signals the same cells in the brain that respond to leptin, but it increases appetite. It also affects our reward system and is part of what makes eating such a pleasant experience. Clearly, both of these hormones play important roles in when and how much we eat.

The Enteric Nervous System: Our Second Brain

Next up in our refresher course is the final piece of the metabolic puzzle—the enteric nervous system, also known as your gastrointestinal (or GI) brain. Now don't go getting all excited: this second brain isn't about to help solve the Sunday crossword puzzle or write a concerto, but it is valuable nonetheless. Hidden in the walls of your digestive system, the GI brain has about the same number of neurons as a cat's brain (and if you've spent any time on the internet, you know all the clever things cats can do) and is essential to how our bodies function and how we feel.

The GI brain consists of a network of about 100 million interconnected nerve cells and clusters of nerve cells that can work independently to regulate gut activity without input from the central nervous system. This network has its own sensory receptors and reflexes, which allows us to go about our day while the GI brain looks after all things digestive. About 90 percent of these neurons carry information to the brain in our head. The other 10 percent go in the opposite direction, carrying information from our central nervous system to our GI brain. This 10 percent works in the background, the same way as the nerves that control heart function and breathing, leaving us blissfully unaware of the process.

About half of the information flow *within* our GI tract is between our neurons and our gut bacteria, or microbiota. This ongoing communication allows the bacteria in our gut to influence how we perceive the world. This can manifest itself as a kind of intuition, or "gut feeling." Have you ever experienced anxiety or excitement as "butterflies in your stomach," or described a traumatic event as feeling as though you'd been "punched in the gut"? That's your GI brain at work. And if it can have such a pronounced effect on our behavior and the way we feel, it goes without saying that having a healthy gut is crucial.

A Fantastic Voyage

When I was a boy, I loved science fiction; still do. One of my favorite movies was *Fantastic Voyage*, starring Raquel Welch and Donald Pleasence. It's about a scientist with a brain injury that needs immediate attention. The only possible treatment is a new technology that miniaturizes a submarine and its crew, allowing them to be injected into a vein and make their way to the brain to perform life-saving surgery. Crazy, right? But it sure got my attention! I was fascinated by the idea of moving around inside the human body to see how it all worked. Who knows? Maybe I owe my keen interest in nutrition, not to mention my career teaching and learning about the human body, to that movie.

Why am I regaling you with stories of my childhood movie obsessions? Because we can use the concept behind the *Fantastic Voyage* to help us understand why the BioDiet is the best possible way to achieve maximum health. So far in this chapter, you've read about macronutrients, hormones, and the GI brain. But what parts do they play in our physiology? To answer that question, we're going to embark on our own fantastic voyage. Like Raquel and Donald, we'll shrink ourselves down, board a tiny submarine, and follow a single glucose molecule—at first part of a larger starch molecule—as it travels through the human body. Why glucose? Although other nutrients follow a similar path through our digestive system, understanding how we metabolize glucose is important in determining why we gain weight and how we go about losing it.

Like most voyages, however, things can sometimes get a little complicated, but overcoming those challenges is not only part of the fun, it's also essential to understanding human health. Knowledge of the basic structures and functions of the human digestive system is key to the BioDiet, so read on. It's well worth the trip.

A Word before We Embark

Like it or not, we humans are really just advanced worms, equipped with what's called a tube-within-a-tube body plan. Our body, minus our arms and legs, is the outer tube. Our inner tube begins at our mouth, includes our esophagus, stomach, intestines, and ends... well... you know where. But here's where things get really interesting. Despite the fact that most of us think of the digestive tract as part of our *internal* environment, it's not. That hollow part of the tube comprising our stomach, intestines, et cetera, is still *external* to our body. Our internal body is the environment *between* the inner and outer tube (see figure 2.3). Stay with me here; it'll all become clear on our voyage through that tube within a tube.

The Mouth and Esophagus

The digestive process for our glucose/starch molecule begins in the mouth. Mastication, or chewing, breaks down our food into smaller pieces and mixes it with saliva, which contains water, mucus, and a digestive enzyme called amylase, among other things. Once we have sufficiently mashed and mixed it, our tongue moves the food containing our tiny glucose molecule to the back of the throat, which stimulates the swallowing process. This food *bolus* then makes its way down the 10-inch-long esophagus and into the stomach, where muscles in the wall of the digestive tract push it along, much like toothpaste being squeezed through a tube. Valves along the way make sure things move in one direction—most of the time (when we get sick, things can reverse direction...).

The Stomach

The stomach is like a washing machine full of acid. It has valves at both ends that, when closed, create a contained environment. Glands in the stomach wall secrete a solution rich in pepsin, an enzyme that digests proteins. Hydrochloric acid strong enough to eat a hole in your kitchen table activates the pepsin, and three layers of muscle in the stomach walls create a powerful churning motion for more mechanical digestion. Very little is absorbed here, with the notable exceptions of alcohol and fast-acting drugs.

As anyone who's ever overeaten knows, the stomach can increase in size—a lot. It can hold up to a gallon of food if we really push it,

Figure 2.3: The anatomy of the digestive system (note the tube-within-a-tube body plan).

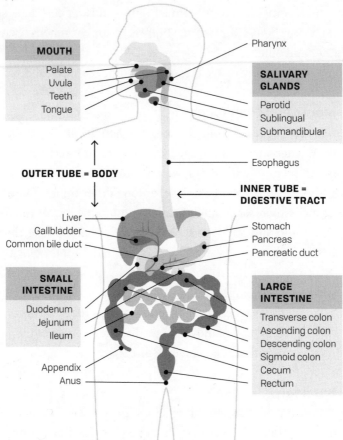

which is why your dad might have loosened his belt after that big Thanksgiving dinner. This expandability was very useful for our ancestral hunter-gatherers who rarely knew when they'd get their next square meal. Food stays in the stomach for anywhere from 20 minutes to two hours, depending on what we've eaten. A valve at the exit point of the stomach regulates the flow of our semi-digested food, called *chyme*, into the small intestine.

The Small Intestine

This roughly 23-foot tube consists of three parts coiled up in our bellies. The first part, the duodenum, is only about six inches long, but it has a vital job to do—receive the chyme from the stomach and mix it with bile from the gallbladder and digestive juices from the pancreas. From there, the food moves on to the second part of the small intestine—the jejunum—an eight-foot-long, thicker-walled coil where the food is further broken down into its smallest sub-units: proteins become amino acids, fats are transformed into triglycerides, and carbohydrates, including our glucose/starch molecule, are turned into monosaccharides.

Freed now from its larger starch molecule, our glucose molecule heads to the third and longest portion of the small intestine, the approximately 12-foot-long ileum, where most absorption occurs. And here's where we get back to the human body's outer and inner tubes. Although vitamins and nutrients in the food can be successfully absorbed from the outer environment of the ileum to the inner environment of the bloodstream, our glucose molecule needs a little help. Enter the glucose transporters (called GLUTs), which facilitate the movement of glucose molecules through cell membranes, including those that separate the gut from the blood vessels.

Before we follow our glucose molecule into the bloodstream, let's find out what happens to the carbohydrates and other organic matter that can't be absorbed into the blood at this stage. This material is headed instead for the large intestine.

The Large Intestine

The main role of the large intestine is to reabsorb into the inner body the nearly two gallons of fluid used daily in the digestive process. It's an important job: without it, we would quickly dehydrate and die. The biggest portion of the large intestine is the colon, which normally hosts a whopping three to five pounds of gut bacteria. These bacteria break down and absorb as-yet-undigested carbohydrates like soluble fiber and resistant starch. They also produce some of our essential vitamins, like vitamin K.

Once all of the useful compounds, vitamins, and water have been absorbed into the body from the large intestine, what's left are feces. These are stored in the rectum until it's convenient to give them their freedom. Enough said.

The Blood Vessels

Back to our glucose molecule, which, thanks to the glucose transporters, has now passed through the wall of the ileum and is headed for the bloodstream. To get a sense of what's in store for us on this part of the trip, it's helpful to know how the body's cardiovascular

The Pancreas

The pancreas is an elongated, tapered organ that sits behind the stomach, tucked in under the liver. It's a gland, which means it secretes substances, but it is somewhat unique in that it is both an exocrine gland *and* an endocrine gland. As an exocrine gland, the pancreas secretes digestive juices, including enzymes, externally via a duct into the small intestine. There, the enzymes break down proteins, carbohydrates, and fats so that the components can be absorbed into the bloodstream and delivered to our cells as fuels and building materials. As an endocrine gland, the pancreas also secretes internally, by discharging hormones, including insulin, into the fluids between cells. These are then absorbed by the blood and circulated throughout the body. This makes the pancreas a two for one, which is why keeping it healthy by watching your weight, getting exercise, and limiting alcohol consumption is so important.

system works. Because the heart is constantly pumping, there's a steady flow of blood from the heart, through arteries, into capillaries, into veins, and back to the heart. (There is also a second circuit—from the heart to and from the lungs—but we don't need to concern ourselves with that.)

The body's arteries and veins get a lot of press, but really they're just highways that carry blood to and from the capillaries. The capillaries are where all the action is. They are the site of important exchanges of nutrients, wastes, and other materials between the blood and the cells. They bring in fuels, building materials, and oxygen, and remove wastes and carbon dioxide. Our glucose molecule is now in a capillary in the gut, dissolved in watery blood plasma. Its next port of call is the liver, where all blood leaving the digestive tract goes first.

The Liver

As blood from the gut passes through capillaries in the liver, the liver regulates what should be in circulation and what should be stored or converted into other compounds. If, for example, the body's blood-sugar levels are low, the liver releases glucose into the bloodstream to be distributed to cells. (Recall that nerve and red blood cells are particularly dependent upon glucose.) High blood sugar, on the other hand, causes the liver to store glucose for later use.

After the skin, the liver is the largest organ in the body, and it has the greatest range of functions—more than 200 in all. Most of these functions are metabolic in nature, converting one compound into another; for example, it turns ammonia into the much less toxic urea and converts fructose and glucose into fat. Other functions include storage of glucose as the complex carbohydrate glycogen; storage of fats and other compounds; packaging lipids for transport; detoxification; production of bile to aid in fat digestion; and the recycling of iron for red blood cells. With responsibilities like that, no wonder some countries, including the US and Australia, have come up with "Love Your Liver" health campaigns.

A very important aspect of the liver's anatomy—and one that is germane to our discussion here—is that the organ has not one but two input blood vessels. The hepatic artery serves oxygenated blood to the liver in the same way that all other tissues are served. But the hepatic portal vein is different. It collects blood from the intestines—blood that contains all the digested and absorbed nutrients that come from our food—and transports it to the liver for processing before it continues into circulation. Because our food contains many more nutrients than we immediately need, the liver acts as a gatekeeper for the rest of the body by absorbing the excess nutrients, like glucose, and letting them trickle out at an appropriate rate between meals.

Cells use two methods for storing glucose: either as glycogen or converted to fat. Glycogen is stored mostly in our liver and muscle cells, but once those are full our bodies convert any additional glucose to fat, which is stored in fat tissue. Unlike glycogen, there is no limit to how much fat we're able to store, which is why some of us grow to be more than 600 pounds. There is a powerful hormonal signal that biases toward storage: the release of insulin. As explained earlier, if insulin is present in normal, moderate amounts, then the glucose in the blood is used for fuel. High insulin levels, however, not only result in the production and storage of fat but can also cause salt and water retention, high blood cholesterol, and increased risk for heart disease, among other problems.

The Cells

So, let's assume that our body's insulin level is moderate, signaling our glucose molecule to enter a cell. There, it can be used in a variety of ways. As a building material, glucose comprises parts of other important molecules both inside and outside of our cells. Glucose can also be stored inside the cell or metabolized to release energy. While most cells have fairly constant fuel demands, some, like active muscle cells, can have enormous, immediate, and persistent energy needs.

If there's a need to metabolize the glucose for energy, our cells have another important consideration. If the energy needs are moderate and sufficient oxygen is available, our molecule will take the aerobic pathway (*aerobic* means "with oxygen"), which uses oxygen to metabolize our glucose molecule more slowly and efficiently, producing only water and carbon dioxide as by-products. If, on the other hand, the energy is required ASAP and/or no oxygen is available, our glucose molecule will find itself on the anaerobic energy

Resistant Starch

Resistant starch is made up mostly of a specific type of carbohydrate called *amylose*. Note the similarity to *amylase*, the enzyme that breaks down carbohydrates in the mouth and small intestine. (If a name ends in *-ase*, it's an enzyme, and if it ends in *-ose*, it's a carbohydrate.)

Amylose is resistant to digestion in the small intestine because its rod-like structure means it piles up like cordwood. This prevents the enzymes in the small intestine from accessing it. When amylose reaches the large intestine, however, bacteria are able to break it down. But amylose isn't reduced to glucose, like other starches. Instead, it breaks down into fatty acids that are absorbed into the blood and have no effect on blood sugar or insulin secretion. The same goes for soluble fiber, which is also broken down by bacteria in the gut into absorbable fatty acids, which in turn stimulate blood flow to the colon, increase nutrient circulation, help us to absorb minerals, and promote the growth of good bacteria in our bellies.

Our bodies are only as good as the fuel we put in them—and unlike cars, there's no opportunity for trading up.

pathway (*anaerobic* means "without oxygen"), which metabolizes glucose quickly but inefficiently and produces wastes that build up fast. This is an important distinction; as we'll discuss later, cancer cells are highly anaerobic.

So there you have it—the fantastic voyage of a glucose molecule through the metabolic process. As you can see, the human body is an incredibly complex system, not unlike a high-end automobile. Give it cheap gas and it isn't long before things start to go awry. In this analogy (and I'm a car guy, so you've gotta let me run with this), carbohydrates are the low-octane fuel that causes engine knock and poor performance. The BioDiet, however, provides the high-octane stuff that is specifically designed to keep things running smoothly and all engine parts clean. Now, some will argue that there's no harm in fueling up with cheap gas, but I've noticed that these are the same people who tell you to ignore the "check engine" light. The truth is that our bodies are only as good as the fuel we put in them—and unlike cars, there's no opportunity for trading up. Ignore the warning light for long enough, and you'll end up broken down on the side of the road. What does that look like? Turns out, we already know.

3

Broken Down
The Axis of Illness

*The mind that opens to a new idea
never returns to its normal size.*
**ALBERT EINSTEIN, THEORETICAL PHYSICIST,
1879-1955**

BACK IN 2002, President George W. Bush first used a pretty evocative phrase: during his State of the Union address, he referred to Iran, Iraq, and North Korea—three countries he considered to be common enemies of the American people—as the "Axis of Evil." I've used a variation on Bush's phrase as the subtitle for this chapter because it perfectly describes the three common enemies of our health and well-being: obesity, insulin resistance, and inflammation.

At the core of the Axis of Illness is a weapon of mass destruction: the high-carbohydrate diet most of us have been eating our entire lives. The more carbohydrates we consume, the worse our obesity, insulin resistance, and inflammation become. But the damage doesn't end there. As these conditions worsen, they feed upon one another: obesity contributes to insulin resistance, which in

Figure 3.1: The Axis of Illness.

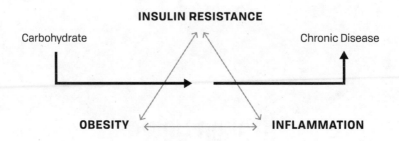

turn worsens obesity; inflammation aggravates obesity, promoting more inflammation; insulin resistance causes more inflammation, which worsens insulin resistance. This vicious cycle (what we call a positive feedback loop in physiology) upsets our internal balance in much the same way as gassing up a car with the wrong kind of fuel. In the case of our health, though, the results can be deadly, made even more so because insulin resistance and inflammation are largely invisible; they slowly worsen and we only become aware when the damage is done.

However, the good news—actually, the *great* news—is that by reducing carbohydrate consumption, the BioDiet combats obesity, insulin resistance, and inflammation. To understand how, I'll borrow a principle from the ancient Chinese military treatise *The Art of War*: "know the enemy"—or, in this case, all three of them.

Obesity

It's a sad fact that the only animals that suffer from obesity are humans and the pets that we feed. But obesity hasn't always been the epidemic it is now. In 1960, just 14 percent of American adults were obese, according to the US Centers for Disease Control and Prevention. That number remained relatively stable until 1980, when the US Departments of Agriculture and Health released the first *Dietary Guidelines for Americans*, in which they recommended a diet low in fat and high in carbohydrates. Twenty years later, that number had risen to 30 percent. Today, more than 40 percent of American adults are obese, and more than 300,000 die from obesity-related causes each year.

In Canada, the rise in obesity rates is slower, but the trend is clear. According to Statistics Canada, obese Canadians represented just 14 percent of the population in 1978, but by 2015 the number had risen to 27 percent. If we include the overweight, that number grows to 62 percent of Canadians. It is a problem that is international in scope and shows little sign of abating. According to the Organisation for Economic Co-operation and Development, 47 percent of Americans, 39 percent of Mexicans, and 35 percent of Brits will be obese by 2030.

Medically, we define obesity by body mass index (BMI), a number popularized by none other than Ancel Keys, the man ultimately responsible for those misguided 1980 food recommendations. BMI is a measure of "appropriate weight" for a given height. If you're curious about your own BMI, I've included the formula for calculating it in chapter 5. If your number is under 25, you are a healthy weight. Between 25 and 30, you're considered overweight. Above 30, and you're considered obese. Over 40, morbidly obese.

For some time now, governments and organizations have attempted to stem North America's obesity tide by urging us to reduce calorie consumption and increase the amount of exercise we get. The overall message is that individual gluttony and laziness are responsible for the collective mess we're in. But a rapidly growing number of others—myself among them—believe in the science, which strongly points to excess *carbohydrate* consumption as the biggest contributor not only to obesity but, as you'll see, to the two other important components of the Axis of Illness.

Diseases Caused or Aggravated by Obesity

Cardiovascular diseases:
- stroke
- atherosclerosis (hardening of the arteries; contributes to heart attacks and strokes)
- hypertension (high blood pressure)
- phlebitis (venous blockage)
- elevated triglycerides in the blood
- low HDL ("good" cholesterol)

Other metabolic diseases and complications:
- cancers (esophagus, pancreas, kidney, colon, cervix, breast, uterus, prostate)
- diabetes
- metabolic syndrome
- non-alcoholic fatty liver disease

Other diseases and conditions:
- cataracts
- gout
- pancreatitis
- osteoarthritis
- infertility
- polycystic ovarian disease
- sleep apnea
- GERD (gastroesophageal reflux disease)

Figure 3.2: The rise in prevalence of obesity in the US coincides with the first dietary guidelines. (*Source: US Centers for Disease Control, National Center for Health Statistics; dotted line indicates OECD projected prevalence.*)

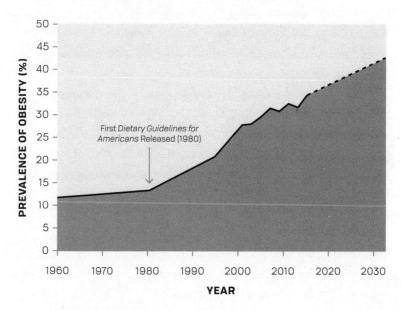

Insulin Resistance

As we learned in chapter 2, the hormone insulin is vital for our survival. It has two principal functions: to facilitate the movement of glucose out of the blood and into body cells, and to help to construct proteins (hence its abuse by body builders looking to bulk up). A third function, to promote the re-uptake of sodium from the kidneys so it remains in our body, also has an important impact on our health. Before we explore what happens when we become insulin resistant, let's first revisit how our bodies regulate blood glucose.

During digestion, foods that contain carbohydrates are broken down into glucose molecules that are absorbed into our blood. This causes a rise in blood glucose, which signals our pancreas to

Determining Blood-Sugar Levels

If you see your physician once a year for a checkup, you're probably no stranger to a blood-glucose test, particularly if you're over 45, have a family history of strokes or hypertension, have high triglycerides or low "good" cholesterol levels, or are overweight or sedentary. All of these can put you at a greater risk of developing diabetes.

When your physician measures your blood sugar, they will ask you to fast for at least eight hours prior to having your blood taken. This is because, after meals containing carbohydrates, your blood-sugar level rises to about 200 mg/dL (milligrams per deciliter) or approximately double its normal peak value. Fasting establishes a baseline for your normal blood-sugar level without the surge that occurs after meals.

If your fasting blood-glucose level is higher than desirable, your physician will likely follow up with a glucose tolerance test. This starts by measuring your fasting glucose via a blood test. Next, you drink a specific amount of a glucose solution and wait two hours before taking a second blood test to see how well your body responds to that glucose challenge. A high blood-glucose level at the second reading indicates impaired glucose tolerance and is the first sign of diabetes. Another test—called hemoglobin A1c—measures the average amount of glucose in the blood for the past few months by estimating how much is attached to the hemoglobin in your red blood cells. These cells only live for about three to four months, and they tend to bind excess blood glucose, making this a practical way to evaluate how much glucose persists in your blood over time.

produce insulin. Insulin's role is to tell the body cells—especially the liver and muscle cells—to absorb glucose from the blood. To use a radio analogy, the insulin effectively commandeers the body's emergency broadcast system to send out a message: *Remove the glucose from the bloodstream.* As the glucose is absorbed, blood-sugar levels return to normal.

Blood-glucose regulation is a delicate balancing act, and, depending on what we eat, we can tip it toward ill health. A diet chronically high in carbohydrates causes chronically high levels of insulin in the bloodstream, which over time causes cells to become less responsive to the signal telling them to uptake glucose from the blood. To return to our radio analogy, it's as if the cells have slowly gone deaf to the emergency message (*Remove the glucose from the bloodstream*). In response, our pancreas "turns up the volume" by producing more insulin. Since high insulin also causes our bodies to store glucose as fat, this can lead to obesity, which results in more insulin resistance. It's a vicious cycle that promotes impaired glucose tolerance, or pre-diabetes, which can lead to full-blown diabetes. Left untreated, this causes damage to the heart, blood vessels, kidneys, eyes, and nerves and will eventually result in premature death.

According to the US Centers for Disease Control and Prevention, today almost half of adult Americans are pre-diabetic (34 percent) or diabetic (10 percent), three times the rate a half century ago. A 2016 US National Health and Nutrition Examination Survey found that 11.5 percent of American deaths were due to complications of diabetes. That's four times the number recorded on death certificates and makes diabetes, in fact, the third leading cause of death after cardiovascular disease and cancer.

The number of Canadians diagnosed with diabetes has grown by an alarming 50 percent over the past 10 years, according to Statistics Canada. Today, 3.4 million (nearly 1 in 10 people) suffer from the disease, and another 5.7 million are pre-diabetic. If the current trend continues, an estimated 44 percent of Canadians will be diabetic by 2025. The numbers are frightening, as are the nasty complications associated with the disease. People with diabetes are 25 times more prone to vision loss, 20 times more likely to have a limb amputated, and 12 times more likely to be hospitalized with kidney failure. Those with type 2 diabetes are also at a 60 percent greater risk of developing dementia.

Insulin resistance (and its resulting high blood-glucose levels) is also a major contributor to cancer, since most cancer cells are dependent on glucose as a fuel: something known as the Warburg effect. The rate of glucose uptake in cancer cells is 100 to 200 times that of normal cells, so high glucose means ample fuel for tumor

A 2016 US National Health and Nutrition Examination Survey found that 11.5 percent of American deaths were due to complications of diabetes.

growth. Furthermore, inflammation and insulin also promote tumor growth and proliferation, meaning that a high-carbohydrate diet provides the fuel and fertilizer that creates an optimal environment for cancer progression.

Diseases Caused or Aggravated by Insulin Resistance

Cardiovascular diseases:

- atherosclerosis (hardening of the arteries; contributes to heart attacks and strokes)
- hypertension (high blood pressure)
- cardiomyopathy (disease of the heart muscle)
- increased blood-clotting risk
- elevated triglycerides in the blood
- low HDL ("good" cholesterol)

Other diseases:

- diabetes
- increased intra-abdominal fat (obesity)
- inflammation and cirrhosis of the liver (hepatitis and degeneration, respectively)
- PCOS (polycystic ovarian syndrome)
- sleep apnea
- acne

Inflammation

Inflammation is an immune response: our bodies' way of trying to protect and heal itself. *Acute* inflammation is a beneficial response to an immediate threat, such as a physical injury or an infection. Small arteries dilate to supply more blood to the damaged region, fluids and proteins move between blood and cells, and the body produces white blood cells that migrate to the area. Damaged tissues are repaired and, if all goes well, any infection is contained and resolved. Acute inflammation starts rapidly, can become severe quickly, and is normally present for just a few days.

Chronic inflammation (from the Greek word *chronos* meaning "time") refers to inflammation that persists for months, even years. It can be caused by an autoimmune disease, by repeated exposure to industrial chemicals or air pollution, or by lifestyle factors such as diet. Some chronic inflammation affects specific tissues and organs, as is the case with osteoarthritis and inflammatory bowel disease. However, when inflammation persists at a low level throughout the body, we refer to it as *systemic* inflammation, which over time can cause significant damage to all cells and tissues. Ironically, the culprits of systemic inflammation are the same white blood cells that rush to the defense of damaged tissues as part of acute inflammation. In the case of systemic inflammation, however, the white blood cells arrive and then hang around, even when the threat no longer exists or didn't require an inflammatory response in the first place. Like unwelcome houseguests, the longer they stay, the more mischief they get up to.

Systemic inflammation can manifest itself as aches and pains or a general feeling of ill health. Aggravated by poor fitness and obesity, chronic systemic inflammation is strongly linked to cardiovascular disease, cancer, type 2 diabetes, and Alzheimer's, sometimes referred to as type 3 diabetes. As brain cells become insulin resistant, they process glucose less effectively, leading to impaired brain function. Chronic inflammation causes further damage by weakening the blood-brain barrier in older adults, which allows pathogens to enter the brain, triggering a buildup of plaque, which results in further inflammation.

Diseases Caused or Aggravated by Chronic Systemic Inflammation

Cardiovascular diseases:

- stroke
- atherosclerosis (hardening of the arteries; contributes to heart attacks and strokes)
- hypertension (high blood pressure)
- cardiomyopathy (disease of the heart muscle)
- heart failure

Other metabolic diseases and complications:

- cancers (lung, kidney, colon, pancreatic, lymphoma, breast, prostate)
- diabetes
- metabolic syndrome
- non-alcoholic fatty liver disease
- chronic fatigue syndrome

Inflammatory diseases:

- irritable bowel syndrome
- chronic obstructive pulmonary disease
- pancreatitis
- psoriasis
- rheumatoid arthritis

Neurological disorders:

- Alzheimer's disease (and other dementias)
- Huntington's chorea
- Parkinson's disease
- peripheral neuropathies

Mental health disorders:

- clinical depression
- bipolar disorder
- schizophrenia

Inflamm-aging

Whenever I talk to people about the Axis of Illness, most are surprised at the considerable negative impact that obesity, insulin resistance, and inflammation can have on our health. But if I really want them to sit up and pay attention, I just have to mention the word "aging"—as in *inflamm-aging*. Because let's face it: even those among us who don't mind getting older don't want to actually "age." I mean, really, haven't we found a cure for that yet? The answer is maybe. Or at least we can likely slow the process.

The term inflamm-aging was coined around 2006 by Dr. Claudio Franceschi of the University of Bologna and Dr. Judith Campisi of the Lawrence Berkeley National Laboratory in California. It encompasses one of today's most widely accepted theories, which is that aging involves compounds called *advanced glycation end-products*—or, aptly, AGEs. *Glycation* refers to the attachment of blood-borne sugar molecules (glucose) to other structural and functional molecules, especially proteins, in our cells and body fluids. These attachments, called cross-linkages, gum up normal molecules, making them less structurally sound and/or inhibiting their normal function. Take skin, for example. As the collagen in our skin—a triple-coiled, rope-like protein—becomes glycated, the skin becomes tougher, less elastic, and wrinkled. That process is AGEs in action, and it's going on inside our bodies, too. Our kidneys, heart, liver, muscles, and even our brain all lose functionality.

It should come as no surprise then that AGEs are suspected contributors to chronic diseases like diabetes, kidney and heart diseases, stroke, arthritis, and Alzheimer's. Google "AGEs" and you'll likely find references to their association with diets high in animal fats, but that's not the full story. While animal fats can become glycated (which is essentially what happens when we cook

The rate of glucose uptake in cancer cells is 100 to 200 times that of normal cells, so high glucose means ample fuel for tumor growth.

them), this does not necessarily cause AGEs to accumulate in the blood and tissues. For that to happen, we need high levels of fat circulating in our blood as well as high levels of blood sugar, both of which are by-products of a carbohydrate-rich diet. By adopting a diet that reduces carbohydrates and fats in the blood—like the Bio-Diet—we're minimizing AGEs, which should help to slow the aging process and further diminish our risk of chronic disease.

Now for the Good News

You've spent the last several pages taking in the bad news regarding the Axis of Illness and the debilitating diseases that result from obesity, insulin resistance, and inflammation—all of which result from a high-carbohydrate, low-fat diet. Now, here's the good news: the BioDiet can help. In the next chapter, we'll go into much more detail about how, but here's a bit of a sneak peek:

- Although the BioDiet isn't a weight-loss plan, per se, it does work very effectively and will result in fat loss, especially around the waist. Over time, the diet will reverse weight gain and obesity, along with the inflammation and enormous complications that arise as a result.

- The BioDiet is designed to address insulin resistance. By severely restricting sugars and starches, your body produces glucose only as needed, and your blood-sugar levels remain normal and

Hypoglycemia and Hyperglycemia

The most recent evidence indicates that the normal range for blood glucose lies between about 80 and 110 mg/dL. If the level drops below 80 mg/dL, we get hungry. We eat, and the food is digested and absorbed into the blood. Because some of this digested food contains glucose, it enables our blood sugar to return to normal. If we don't eat, we soon start to feel tired, sleepy, and maybe a little cranky. *Hypoglycemia* (literally "too little glucose in the blood") occurs if the blood-glucose level drops below 40 mg/dL. At that point, more serious symptoms, like mental confusion and significant muscle weakness, begin to manifest. A blood-sugar level below 15 mg/dL is likely to result in seizures and coma. Death soon follows.

Elevated blood sugar, or *hyperglycemia*, is not as serious a problem, at least in the short term. The generally accepted blood-sugar level for hyperglycemia is above 180 mg/dL, though it's not until levels reach the 250 to 300 mg/dL range that symptoms begin to appear. These include excess urination (which can lead to thirst and dehydration), headaches, blurred vision, and trouble concentrating. Persistent hyperglycemia leads to the complications of diabetes described on page 82.

steady. This decreases the need for insulin by roughly 50 percent and quickly increases your cells' sensitivity to it. By directly addressing the root cause of insulin resistance—persistent, excess carbohydrate consumption—the BioDiet reduces the likelihood of developing chronic diseases by about 70 percent overall (this is a rough estimate).

- Insulin resistance is also a major contributor to aging. The creeping increase in blood glucose due to insulin resistance causes an

Free Radical Danger

Another important factor that contributes to inflammation is DNA damage caused by reactive oxygen species (ROS), or free radicals, but you're likely more familiar with the term oxidants, for which we might take *anti*oxidants, like vitamin C, as supplements. Most oxidants are produced in the mitochondria, which are cell structures responsible for efficiently combining fuel from the food we eat and oxygen from respiration to release energy, without which we'd die. As the mitochondria metabolize glucose, they produce inflammatory oxidants resulting in more DNA damage. Oxidants are especially harmful to those body cells that don't reproduce, like nerve and muscle cells, which are exposed to accumulating oxidants for longer periods, making them more likely to suffer DNA damage.

Although our cells are capable of reversing much of this DNA damage, they are less able to keep up as we age (a comparison of cells in the very young and the very old reveals 5 to 10 times more damage in the older cells). If a cell becomes so damaged that it dies, it can burst, releasing a host of inflammatory chemicals into circulation, which then travel throughout the body. This further aggravates systemic inflammation, creating a damaging cycle that can lead to significant chronic disease.

Chronic systemic inflammation is strongly linked to cardiovascular disease, cancer, type 2 diabetes, and Alzheimer's.

increase in the production of AGEs, which promote systemic inflammation. By restricting carbohydrates, the BioDiet not only limits blood glucose, but it also reduces triglycerides in the blood by as much as 70 percent, which means less sugar and fat to fuel AGEs. This alone has a very powerful, positive impact on the changes associated with aging.

- The carbohydrate restriction in the BioDiet causes the production of ketone bodies. We'll talk about the beneficial effects of ketones in chapter 4, including their use as a potential therapeutic treatment for chronic disease and cognitive decline.

- Those on the BioDiet use fats as their primary fuel source, and since fats burn more efficiently than glucose, they produce far fewer oxidants. This reduces inflammation and helps to preserve mitochondrial health and therefore the health of the cell. It also results in lower fats in the blood, which helps reduce the risk of cardiovascular disease.

Feeling motivated yet? Keep reading ...

Table 3.1: High-carbohydrate diet vs. BioDiet—an Axis of Illness comparison.

High-Carbohydrate Diet	BioDiet
High blood glucose + high blood triglycerides → increase in AGEs → increased systemic inflammation → premature aging and greater incidence of chronic disease	Low blood glucose + low blood triglycerides → decreased rate of production of AGEs → decreased systemic inflammation → slower aging and less chronic disease
High blood glucose → high glucose utilization in the mitochondria → high levels of oxidants → increased mitochondrial and cell damage → increased systemic inflammation → premature aging	Low blood glucose → high fat utilization in the mitochondria → fewer oxidants → decreased mitochondrial and cell damage → decreased systemic inflammation → slower aging and less chronic disease
Insulin resistance results in high blood glucose	Insulin resistance is reversed and blood glucose returns to normal
Obesity increases systemic inflammation	Obesity and inflammation are reduced

4

The BioDiet Difference
Why It Works

When diet is wrong, medicine is of no use.
When diet is correct, medicine is of no need.
ANCIENT AYURVEDIC PROVERB

I N CHAPTER 3, you learned what happens when you eat a high-carbohydrate diet: in short, it produces an excess of glucose in the blood, which over time leads to obesity, insulin resistance, and inflammation—the root causes of most chronic disease. The best and simplest defense against the Axis of Illness is to restrict carbohydrate consumption. However, our cells still need a steady supply of glucose. If we drastically cut our carbohydrate intake, where will we get that important source of fuel? And if we're increasing our fat consumption to make up for reducing carbohydrates, won't eating all that fat make us fat?

To answer those questions, we need to delve deeper into our understanding of how the body metabolizes the fuels we consume. And ironically the best way to explain the principles of metabolism is to first consider what happens when we stop eating altogether.

> The best and simplest defense against the Axis of Illness is to restrict carbohydrate consumption.

The Glucose Priority

When the body begins to starve, its number one priority is the maintenance of the organ systems central to survival: paramount is the brain, a highly active organ that has little tolerance for shortages of oxygen or fuel. However, unlike most other organs, the brain—along with red blood cells—can't effectively use fatty acids for fuel. That means the body has to find enough glucose, roughly 200 grams per day, to keep all of these components alive. Once the glucose in the blood has been exploited, the body's next step is to tap into the glucose stored as glycogen in the muscle cells and, especially, in the liver.

Releasing Stored Glucose

Glycogen consists of thousands of glucose molecules bound together like Lego blocks into complex, branching chains. When we're between meals or have gone for a long period without eating, our pancreas produces tiny amounts of *glucagon*, a potent hormone that breaks down these building blocks and releases glucose into our blood. The glucose from glycogen in our liver can be released into the blood for all cells, but the muscles are one-way streets: glucose can go in, but it can't come out. That glucose is used internally by working muscles. With a limited supply of glycogen, we can burn through it all in a matter of hours, and even faster when we're exercising. This sounds like a pretty dire situation, and it would be, if it weren't for one nifty trick: once we've used all of the glucose from our glycogen stores, our body cleverly creates its own glucose.

Making New Glucose

The making of new glucose is called *gluconeogenesis* (*gluco*, "glucose," *neo*, "new," and *genesis*, "creation"). It's a process by which your liver and kidneys make glucose from non-carbohydrate sources, including lactate (from lactic acid), glycerol from triglycerides, and amino acids from proteins. Gluconeogenesis helps ensure that even when we're not consuming carbohydrates, our blood-sugar levels don't fall dangerously low (hypoglycemia). It also provides necessary fuel for those tissues obliged to use glucose. This requires only about a teaspoon of glucose (a little more than 4 grams) in the blood at any point in time. As that gets taken up and metabolized in our cells, gluconeogenesis replenishes it.

There are significant by-products of this process, including the amine portion of amino acids, which forms ammonia. Fortunately, our livers are well equipped to deal with this toxic substance. By combining the ammonia with readily available carbon dioxide, the liver creates urea, which is cleared by the kidneys and is excreted in urine. As the urea breaks down it releases the ammonia, which is what gives urine most of its pungent odor.

Gluconeogenesis is not unique to humans. In fact, almost all plants, animals, fungi, bacteria, and other microorganisms have the ability to make their own glucose. In ruminants, like cows, deer, goats, and sheep, gluconeogenesis tends to be an ongoing process. In other animals, as in humans, it takes place during periods of starvation, intense activity, or as a result of low-carbohydrate consumption.

Fat Burning

Your body is pretty obsessed with glucose. In addition to conserving as much of this stuff as possible for those organs that need it most, it is also able to tap into another source of energy for the cells that aren't quite so picky. That source? The fat that many of us are carrying around our waists—and increasingly so, according to the US Centers for Disease Control and Prevention. The CDC reports that the average waist circumference of American adults increased by about 1.2 inches between 1999 and 2011. Some demographic groups, like African American women in their 30s, saw their average waist size grow by nearly 4.6 inches during that time. I can commiserate. At my heaviest, I was carrying about 27 pounds of excess fat, most of it in the spare tire around my torso. I remember sitting slouched in a chair one day and wondering if I could balance a can of beer on my belly. It turned out I could, and that's not a good thing.

As you'll recall from our fantastic voyage in chapter 2, adipose tissue is specifically designed to store a nearly limitless amount of excess energy as fat (triglyceride). We don't produce many more fat cells to accommodate it; the ones we have just keep getting larger, and so do we, which is why we have had TV shows like *The Biggest Loser* and *My 600-lb Life*.

The process of burning fat in these cells is called beta-oxidation, and it has a leg up on carbohydrate burning in a couple of ways.

First, as we discussed in chapter 2, the energy density of fat is more than twice that of carbohydrates, which means it's a more efficient source of energy. For this reason, many athletes, especially endurance athletes, have turned to ketogenic diets. With a ketogenic diet, fat provides about 70 percent of the fuel for the majority of our cells. The other 30 percent continues to come from glucose. A second benefit of beta-oxidation is that, in the absence of carbohydrates, it produces beneficial by-products called ketones. We'll take a closer look at the important functions of ketones in just a moment.

Are there other sources of fuel that your body could tap into—proteins, for example? Technically, yes, but your body is designed to conserve proteins because of their crucial structural and functional roles. It is only once you've depleted all fat stores (during periods of prolonged starvation) that your cells turn to proteins as an energy source. Returning to our cabin analogy, this is akin to breaking up the furniture to provide fuel for the stove. The persistent metabolism of proteins will result in permanent damage to the body and eventually death; obviously, that's something we want to avoid.

Let me be clear: although the BioDiet takes advantage of some of the metabolic processes that occur when we stop eating, it is in *no way* a starvation diet. In fact, it's just the opposite. There are no restrictions on the number of calories you can consume on the Bio-Diet, as long as they're coming from fat, protein, or soluble fiber. Much of the fat you burn will come from the adipose tissue around your waist, which means a slimmer, trimmer waistline. And the fat you consume won't make you fat; instead it will provide valuable fuel for your cells and help to form beneficial ketones. As for the body's precious proteins, you conserve those so they can continue their important structural and functional work.

Ketones: Nature's Superfood

One of the by-products of fat metabolism is what we call ketones. There are actually three ketones that result from the process: acetone (essentially nail polish remover), acetoacetate, and beta-hydroxybutyrate (technically a carboxylic acid, but we'll include it in the ketone group to keep things simple). Acetone is toxic but, like urea, is quickly and easily removed from circulation by the kidneys and eliminated in the urine or exhaled in the breath. The other two ketones, acetoacetate and beta-hydroxybutyrate (BHB), are burned as fuels, primarily by the heart and brain.

BHB in particular is a highly beneficial molecule that's receiving a lot of attention these days in the scientific community. In addition to being a preferred fuel and a potent cell-signaling molecule akin to a hormone, BHB does some pretty amazing things at the cellular level, making it an effective therapeutic for treating and reducing many chronic diseases. We'll discuss this further, but for now it's good to keep in mind that diets that are merely low-carbohydrate

A Word about Ketoacidosis

The production of ketones that occurs on the BioDiet and other ketogenic diets is completely normal. For this reason, Dr. Stephen Phinney, now the chief medical officer for Virta Health, coined the term *nutritional ketosis* to differentiate it from the extremely elevated level of blood ketones called *ketoacidosis*. Often the result of severe diabetes, ketoacidosis is the dangerous buildup of acids in the blood caused by levels of ketones two or more times greater than those produced in nutritional ketosis. Although the similarity in terms can spark concern among some health professionals, neither you nor they have reason to be worried.

Think of ketones as a kind of high-octane fuel for your body.

but not low enough to produce ketones don't have nearly the same benefits as those that do. By helping you to ramp up your production of BHB, the BioDiet can optimize your heart and brain functions, activate anti-aging genes, boost fat loss, and improve your overall health and lifespan.

Metabolizing Ketones

Like macronutrients, ketones (save for acetone) are metabolized as fuels in our cells. However, compared to glucose (from carbohydrates), ketone bodies are like fat: they burn more efficiently, produce fewer oxidants, and result in less inflammation. Think of ketones as a kind of high-octane fuel for your body. Organs like muscles and the heart can run on lower-octane fuel, including glucose and lactic acids, but, like cars, their best performance comes from energy-dense ketones and fatty acids. And since the heart is really the one thing we can't live without, it's wise to give it what it wants when you're gassing up.

When I was a university undergrad in the late 1970s, we were taught that the brain could be fueled *only* by glucose. We now know that this isn't true. Though the brain does need glucose, it can also metabolize ketones. In fact, the brain prefers BHB, which is why within the first few days of a non-carbohydrate diet, the brain will derive about 25 percent of its energy from BHB. What's more, the brain, which consists mostly of fatty tissue, uses BHB to produce lipids for the brain cells' internal functions and repair.

Some speculate that the brain actually works better when it's fueled by ketones instead of glucose, although this has yet to be proven conclusively. When you consider, however, that ketone levels can reach as high as 70 percent during periods of prolonged starvation, this hypothesis makes a certain amount of sense: surely we'd want to be at our sharpest when our very survival is at stake. As an aside, almost all of the people I've counseled on the BioDiet report a newfound mental clarity, often described as having a "fog" lifted from their brain—a sensation I also experienced.

The BioDiet and Your Health

Though nutritional ketosis is at the foundation of the BioDiet, there is much more to it. In part 2 of this book, I'll take you—step by step— through the process of embracing this much healthier way of eating. But before we get to that, I'd like to spend some time exploring how the BioDiet can have a positive impact on your health in ways that don't necessarily show up on the bathroom scale. Let's start with a recap of what occurs when we eat a high-carbohydrate diet.

Elevated carbohydrate consumption raises our blood glucose, causing our pancreas to secrete high levels of insulin, which over time impairs our cells' ability to take up and metabolize glucose.

The BioDiet Difference: A Summary

- The restriction of non-fiber carbohydrates stimulates fat metabolism and ketosis, fat-burning processes that result in fat loss, especially in the mid-abdominal region.

- Fat metabolism and ketogenesis ensure that fat in the diet doesn't stay in the blood and therefore has no detrimental effect on health.

- The metabolism of fat reduces excess weight and lessens obesity.

- Increased fat consumption reduces hunger between meals, provides more effective satiety (feeling full), and prevents overeating.

- The restriction of non-fiber carbohydrates also helps to moderate blood sugar, keeping it constantly within a normal range, which reduces the over-secretion of insulin that leads to insulin resistance.

- Ketones help to reverse insulin resistance and activate genes that resist oxidative stress; they also burn efficiently as a fuel, especially in the brain.

- Systemic inflammation is reduced in concert with decreases in body fat (obesity) and insulin resistance and through the metabolism of ketones.

- All three components of the Axis of Illness—obesity, insulin resistance, and inflammation—are reduced, disease symptoms decrease, and your body returns to its natural, healthy state.

Figure 4.1: How insulin resistance causes chronic disease.

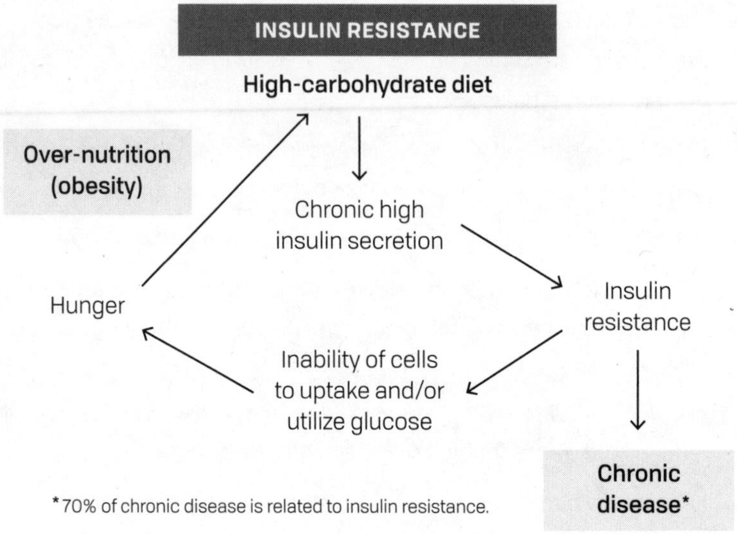

This has several important effects:

- dysfunctional glucose metabolism causes us to be perpetually hungry, so we overeat, which contributes to obesity;
- the chronic excess glucose in our blood leads to insulin resistance and, for a growing number of us, diabetes; and
- the combination of obesity and insulin resistance causes inflammation—a nasty trifecta that, left untreated, will make us sick and shorten our lifespan.

It's a stark scenario but far from inevitable. We have the power to change the trajectory of our health and well-being simply by changing what we eat. The BioDiet will help fend off, treat, and, in some cases, reverse chronic conditions, all while helping you to establish a healthier, tastier way of eating. Let's break it down.

Cardiovascular Disease

Cardiovascular diseases, including heart attacks, strokes, and other circulatory diseases, are the leading cause of premature death globally, killing 17.3 million people in 2013 alone, up from 12.3 million in 1990. More than half of that increase can be attributed to ischemic heart disease, also known as coronary artery disease. Most of the remainder is due to strokes. These are conditions in which the major blood vessels that supply our hearts and brains with oxygen and nutrients become stiff and narrow. Often the first sign is high blood pressure, but when left untreated or poorly controlled, this can lead to more imminent, life-threatening events like heart attacks and strokes.

As we've seen, fat and cholesterol in the diet have long been thought to be the cause of coronary artery disease, but scientific evidence increasingly points instead to inflammation as the primary

We have the power
to change the trajectory
of our health and well-being
simply by changing
what we eat.

Fat and cholesterol in
the diet have long been
thought to be the cause
of coronary artery disease,
but scientific evidence
increasingly points instead
to inflammation as
the primary culprit.

culprit. Here's how it works: oxidants and excess sugar in the blood trigger inflammation, causing arteries to become laden with fatty deposits called plaque; the body perceives the plaque as an intruder and sends in white blood cells to attack, resulting in further inflammation and eventually leading to atherosclerosis. The arteries become stiffer and narrower, causing the heart to work harder to push sufficient blood through the compromised vessels; pressure can increase until it results in hypertension, which is characterized by resting blood pressure in excess of 130 over 80 (millimeters of mercury). In an effort to control the pressure, the vessels harden even further and can eventually rupture. Should a piece of plaque break free, it can lodge itself in the narrowed vessels or trigger a blood clot, reducing blood flow even further. Blockage of an artery to the heart can cause a heart attack, while a blocked artery in the brain can result in stroke.

The BioDiet helps to prevent and even reverse cardiovascular diseases though several different means: by reducing carbohydrate consumption, inflammation caused by high blood sugar is significantly curtailed; arteries don't become inflamed, which reduces plaque buildup; stiffening and narrowing vessels and the accompanying hypertension are avoided; and fewer fatty deposits diminish the risk of blood clot formation and lessen the opportunity for a piece of plaque to break free and wreak havoc.

In addition, the ketones that are a by-product of the BioDiet have their own direct anti-inflammatory properties, while the decrease in body mass that results from a ketogenic diet quite literally lightens the load for the heart. And the BioDiet typically increases HDL, the "good cholesterol," which reverses atherosclerosis. Taken together, these factors greatly reduce our risk of heart attack, stroke, and other circulatory diseases.

Diabetes

As we learned in chapter 3, diabetes has reached epidemic proportions in both the United States and Canada, and pre-diabetes, which left untreated often leads to full-blown diabetes, has skyrocketed. Australia has seen similar trends. The unprecedented rate at which these numbers have risen led World Health Organization director general Margaret Chan to describe the epidemics of diabetes and obesity as "a slow-motion disaster."

Our exploration of the Axis of Illness revealed that type 2 diabetes causes or worsens both obesity and inflammation and directly contributes to other chronic conditions including heart and blood vessel damage; nerve, kidney, and eye problems; and Alzheimer's. Diabetes also poses a significant and growing economic burden. According to a study published in *The Lancet: Diabetes and Endocrinology*, the estimated global cost of diabetes in 2015 was $1.31 trillion. In America alone, it is expected that, by 2030, the total economic burden of diabetes will reach $622 billion, roughly equivalent to the 2018 US defense budget. Clearly, getting a handle on this costly and life-threatening disease is crucial.

The BioDiet not only reduces the incidence of insulin resistance and diabetes, it can help reverse them. It does this in a variety of ways. By eliminating fructose-containing sugars, the BioDiet puts a halt to fructose-induced lipogenesis (fat production) that occurs in the liver; improves insulin sensitivity by as much as 75 percent in as little as a month; and keeps blood-glucose levels within the normal range, so less insulin is secreted and less fat is deposited in the adipose tissue as a result. By restricting non-fiber carbohydrates, the BioDiet also triggers the liver to produce glucose via gluconeogenesis as needed; allows the body to shift from burning glucose to burning fat, thereby reducing triglyceride levels; and helps prevent overeating through favorable changes to the hormones leptin and ghrelin, which regulate hunger and satiety.

Almost all of those
on medications for high
blood sugar, hypertension,
or chronic inflammatory
conditions were able
to reduce or, in some
cases, completely
eliminate their need for
prescription drugs.

In my counseling of those who've adopted the BioDiet, I've seen some remarkable transformations. Some subjects lost more than 30 pounds within the first three months, although more typical was a loss of 10 to 15 pounds for women and 20 to 25 pounds for men. Perhaps more importantly, almost all of those on medications for high blood sugar, hypertension, or chronic inflammatory conditions were able to reduce or, in some cases, completely eliminate their need for

Everything Old Is New Again

Ketogenic diets have been used to treat disease for almost a century. In the 1920s, physicians introduced the ketogenic diet as a treatment for childhood epilepsy. Although we're still not entirely clear on how the diet controls seizures, there is compelling research to indicate that it is the ketone bodies—beta-hydroxybutyrate and especially acetoacetate— that are responsible for the anti-seizure effects. The diet was the standard of care until 1938, when anti-seizure drugs were introduced. In the early 1990s, the diet was re-popularized by researchers and clinicians at Johns Hopkins University for patients unresponsive to drug therapies.

Today, ketogenic nutritional therapy is used in approximately 250 hospitals around the world and is recognized as a standard treatment for some forms of epilepsy. Much of the credit for the diet's resurgence goes to the Charlie Foundation for Ketogenic Therapies, named for Charlie Abrahams (son of American film director Jim Abrahams). In 1993, just shy of his first birthday, Charlie developed a difficult-to-treat form of epilepsy. Despite brain surgery and a punishing drug regimen, the toddler continued to experience up to 100 seizures a day. Physicians had no answers, prescription drugs. Taken together, these facts indicate how successful the BioDiet can be in addressing obesity and diabetes. Of particular importance: during the writing of this book, the diabetes associations in Australia, New Zealand and, most recently, the United States issued policy statements supporting low-carbohydrate diets (including ketogenic diets) as effective measures for the treatment of type 2 diabetes.

so Jim Abrahams went in search of one and found the ketogenic diet. Although, at the time, the little boy's physicians tried to dissuade the family from putting Charlie on the diet, the Abrahams were determined to give it a try. Under the guidance of an experienced ketogenic dietician at Johns Hopkins, they began to feed him a diet rich enough in protein to meet his nutritional needs but very low in carbohydrates. Charlie's seizures began to diminish almost immediately and within a month he was seizure- and medication-free. He remained on the diet for five years. Now, more than 20 years later, Charlie continues to be seizure-free.

Charlie is not alone. Multiple clinical trials have shown that more than half of children treated with a ketogenic diet show significantly reduced incidence of seizures, and about 10 to 15 percent are considered cured. The Charlie Foundation (CharlieFoundation.org) continues to support research into the therapeutic benefits of nutritional ketosis for the treatment of a range of diseases and is a valuable resource for those looking to adopt a ketogenic diet.

Alzheimer's Disease: Type 3 Diabetes?

In 2005, Dr. Suzanne de la Monte of Brown University coined the term "type 3 diabetes" in reference to Alzheimer's disease. Research since then supports this notion, and although the label is still debated, it's now much more widely accepted.

A quick breakdown of the various types of diabetes helps to put things into perspective. When we talk about "diabetes," we are often using the general term to refer to type 2 diabetes mellitus, because that condition comprises roughly 90 percent of all cases. Type 2 diabetes is due primarily to insulin resistance. Type 1 diabetes, on the other hand, is characterized by a lack of insulin production due to an autoimmune attack on the cells that produce insulin, found in the pancreas. There are a few other types as well, such as gestational diabetes, which sometimes occurs during pregnancy. All forms of diabetes are defined as chronic elevated blood-glucose levels resulting from impaired glucose metabolism due to insulin resistance (type 2) or poor insulin production (type 1). According to de la Monte, Alzheimer's disease is essentially insulin resistance of the brain and therefore constitutes a third, discrete variety: type 3.

Alzheimer's is the most common form of dementia, and the second most common neurological disorder after strokes. A chronic, degenerative condition, Alzheimer's results from the progressive destruction of the neurons in the brain. It is characterized by memory

loss, cognitive decline, and premature death and can only be defin-
itively determined postmortem by the presence of structures called
neurofibrillary tangles and deposits of a protein called beta amyloid in
the brain and its blood vessels.

As discussed in chapter 3, insulin is necessary for cells in the fat
tissue, heart, and muscles to transport glucose from the blood into
the cells where it can be stored or metabolized. However, the brain
has different glucose transporters that don't require insulin. So, how
does the brain become insulin resistant? Here's where it gets a little
more complicated. Although the brain doesn't require insulin for the
transportation and uptake of glucose, it does need it for brain cells to
effectively use the glucose. If you think of the brain as an unlocked car,
you don't need a key (insulin) to get into the vehicle, but you do need it
to start the ignition and get the gas (glucose) to the car's motor. As the
brain becomes insulin resistant, the key may still work, but less gas
flows to the engine.

Okay, so the analogy falls apart a bit at the end there, but you
get my point. A lack of brain insulin (or similarly structured signaling
chemicals, aptly called insulin-like growth factors) could be the root
cause of the biochemical and pathological changes we observe in
Alzheimer's disease.

Neurological Diseases

We know that ketogenic diets can work well for some with epilepsy, but what about those suffering from the most common type of neurodegenerative disorder—dementia, including Alzheimer's disease and vascular dementia? As anyone who's ever witnessed a loved one suffering from dementia knows, this is a tragic condition characterized by failing memory, declining communication and language skills, impaired judgment, and deteriorating social and emotional control.

What many might not know is that Alzheimer's—which is named for German physician Dr. Alois Alzheimer, who identified the disease in 1906—is increasingly referred to as type 3 diabetes (see page 86, "Alzheimer's Disease: Type 3 Diabetes?"). That's because diabetes can impact the brain in a variety of different ways. Insulin resistance results in high blood-sugar levels, which in turn cause inflammation that can reduce or block the flow of blood to the brain. Since many people with type 2 diabetes don't know they have it, the damage can go unnoticed and worsen over time. The risk of type 2 diabetics developing dementia is increased by 60 percent compared to those without diabetes. That additional risk is even greater for women.

By combating insulin resistance, the BioDiet helps to eliminate the inflammation that contributes to reduced blood flow to the brain and increases the brain's capacity to use glucose effectively. The Bio-Diet also produces the brain's preferred fuel, beta-hydroxybutyrate, which when metabolized results in fewer of the oxidative by-products that exacerbate inflammation. Since ketones have also been shown to improve cognitive ability in those with the mild impairment that accompanies aging, it follows that a ketogenic diet will help reduce the worrying decline in memory and thinking skills that many of us experience as we get older. This helps to explain why most people who've adopted the BioDiet report improved mental clarity and a greater ability to focus.

Cancer

I've avoided addressing this scourge of a disease until now. Not only is the subject complex enough to fill volumes, it's also very difficult for me to write about. I was just four years old when my mother was diagnosed with breast cancer. We were a typical Protestant family in the mid-1960s, which meant that we never talked about her illness. In fact, we actively *didn't* talk about it. Her name was Ruth. She was a nurse and one of the bravest people I'll ever know. For years, she endured one painful treatment after another—surgeries, chemo, rudimentary radiation therapy, all with the hope of getting better and perhaps even helping others afflicted with the disease. She died in May 1970 at the age of 46. I was just nine years old. By the time I graduated from high school, I had decided to pursue a career that would allow me to help sick people get healthy. Forty years later, I'm still at it. As the saying goes, the worst thing in your life can sometimes contain the seeds of the best.

Seeds, in fact, are a good place to start when it comes to explaining the impact of a high-carbohydrate diet on the growth and proliferation of cancer cells. Imagine for a moment that you planted seeds in your garden in the spring. By the time the summer rolls around, nature has done its job. Your flowers (healthy cells) have made food (glucose) from sunlight, and fertilizer (insulin) has helped them to flourish. Sure, there are weeds (tumor cells) that pop up here and there, but nothing a trowel (your immune system) can't take care of. As the season progresses, however, you notice that more aggressive weeds (malignant cancer cells) have taken root and are now sucking up the sun, gobbling up the fertilizer, and growing bigger and faster than anything else. So fast, in fact, they threaten to destroy the entire garden (your body).

What's the solution? By restricting carbohydrate we not only limit the amount of sunlight (glucose) the garden gets, we also reduce the amount of fertilizer (insulin). The theory—supported by me and other medical researchers—is that a ketogenic diet helps starve the voracious weeds while still providing enough sun and food to grow your flowers (healthy cells).

I want to be very clear about something: I'm not a physician, and I am in *no way* suggesting that you treat cancer by simply limiting your carbohydrate consumption. I wish it were that easy! What we are hoping, though, is that by combating the Axis of Illness, a ketogenic diet can help improve health outcomes for those suffering from diseases including cancer. Those embarking on a ketogenic diet should do so under the supervision of a physician. This is doubly important for those with cancer or any other chronic condition.

The Warburg Effect

As we learned in chapter 2, normal, healthy cells can use carbohydrates (glucose), fats (fatty acids), or sometimes proteins (amino acids) to meet their energy needs, and they use oxygen to burn these fuels in cellular substructures called mitochondria. We call this aerobic cell metabolism (*aerobic* means "with oxygen"). In cancer cells, however, the mitochondria are dysfunctional, inhibiting the cells' ability to use oxygen to burn fuel. Most cancer cells rely on anaerobic metabolism (meaning "without oxygen"), whereby energy is extracted from glucose through fermentation, an effect first observed by German physiologist and physician Otto Warburg. "Cancer, above all other diseases, has countless secondary causes," he wrote. "But, even for cancer, there is only one prime cause. Summarized in a few words, the prime cause of cancer is the replacement of the respiration of oxygen in normal body cells by a fermentation of sugar."

In short: cancer cells eat glucose.

Warburg was awarded the Nobel Prize in Physiology or Medicine in 1931 for his discovery, which is now known as the Warburg effect. Today, physicians use it to detect cancerous tumors in their patients by taking advantage of the fact that cancer cells are so heavily reliant on glucose for anaerobic metabolism that they uptake 100 to 200 times as much glucose as do healthy cells. Medical physicists tag the glucose molecules with a radioactive isotope of fluorine (producing fluorodeoxyglucose), allowing them to "see" the size and location of the tumors with positron emission tomography (a PET scan).

Balance of Power

The word *tumor* is frightening to us because we associate it with cancer. But a tumor is simply a mass of abnormal cells, and the vast

Treating with Keto?

I'd been researching the health benefits of the ketogenic diet for many years when I was introduced to Dr. Gerald Krystal at the BC Cancer Research Centre in Vancouver in 2015. A distinguished immunologist, Dr. Krystal has been investigating the effects of diet on the risk and incidence of cancer. In 2017, Gerry and I, along with lead investigator Dr. Jeff Volek and his team at Ohio State University, embarked on a three-year study to investigate the therapeutic benefits of a ketogenic diet for women with metastatic breast cancer. At the time of writing, we don't yet know if a ketogenic diet can help improve their health outcomes, but we wouldn't be conducting this trial if we weren't very hopeful. Should it prove successful, we also hope ketogenic diets could accompany the treatment of other forms of cancer, especially glioblastoma multiforme, a deadly type of brain cancer.

majority of them pose no health risk as long they stay where they arise and don't grow so large that they interfere with normal bodily functions. Sometimes we even find benign tumors attractive; think of the small moles we've named "beauty marks."

However, a minority of tumors don't stay put. Instead, they grow uncontrolled until they spread to healthy tissues and organs. The term we use for this is *metastasis*, from Greek meaning "rapid transition from one point to another." Such tumors are called malignant and, because they pose a threat to our health and survival, our bodies have evolved powerful mechanisms to identify and destroy them.

When we are young, our immune systems are highly vigilant and effective at preventing tumor growth. Special white blood cells called macrophages and even more cancer-specific cells called natural killer cells and killer T-cells are on constant alert and will attack and even eat cancer cells. But as we age, our defenses diminish. Our immune systems tire and become less effective. Not only do they have trouble identifying tumors, but they're also less robust, which is why we see such an increase in the incidence of cancers among the elderly. Of course, children get cancers as well. Those forms tend to be very aggressive and have high mortality rates. I believe this is due, in part, to the fact that a child's metabolism is directed at promoting growth and these growth factors allow cancers that do arise to thrive.

In a macabre sense, there is an ongoing arms race between cancer and the immune system. While a ketogenic diet is not a cure for cancer, preliminary research has found that, when used in conjunction with standard treatments, it can help tip the balance in our favor, both by diminishing cancer cells' rates of growth and by improving our immune responses. Because, in order to beat cancer, it is ultimately the immune system that must win the day.

Aging

I mentioned the impact of inflammation on aging in chapter 3, but I'd like to further explore the role that mitochondria play in the process. First, it's important to understand where these structures came from and how they ended up in our cells in the first place. This all happened about two billion years ago, so we weren't there to witness it, but our best guess is that a mitochondria-like bacteria was engulfed by a newly forming cell. The two then established a symbiotic relationship whereby the bacteria received a hospitable environment and plenty of food in return for supplying ample amounts of energy to its much larger host. This relationship allowed cells to advance from bacteria-like organisms to complex life forms, including us.

We now call these cellular substructures mitochondria. What makes them different from other cell components is that they

The Chromosomal Theory of Aging

Another theory of aging involves damaged chromosomes. When we're young, the ends of our chromosomes are capped by protective structures called *telomeres*, which are shaped like the plastic ends of shoelaces (called *aglets*, for you trivia buffs). As our cells divide (and, with the exception of muscle and nerve cells, they do), the telomeres get shorter. Over time, they become so short that they no longer protect the chromosomal material, and the DNA becomes damaged and dysfunctional, like shoelaces that have frayed. We all have an enzyme called *telomerase* that acts to reverse the damage to our DNA, and the more active the telomerase, the more youthful the cells. Recent studies have shown that ketones have a positive effect on telomerase, so they have the potential to slow the aging process.

contain their own DNA, allowing them to reproduce when needed. For example, when we get regular aerobic exercise, say by running, we put an increased demand on our muscle cells to provide more energy for contraction. The mitochondria respond by reproducing within our cells, which increases energy capacity.

Like us, however, mitochondria can become dysfunctional as they age. Our cells have the ability to recognize and destroy failing mitochondria and recycle their components to construct new, high-functioning ones. But sometimes this renewal process goes awry and the still-intact, dysfunctional mitochondria release nasty chemicals into the cells that cause damage to other cell structures and even the nuclear DNA. The cells have some capacity to neutralize small amounts of these chemicals, but when the production exceeds the cell's capacity to cope, the mitochondria become the enemy within, causing cells to become dysfunctional, age prematurely, die, or transform into the abnormal cells we call cancer.

Advanced glycation end-products (AGEs) also contribute to the aging process. Glycation—the attachment of excess blood-borne glucose molecules to otherwise normal proteins and lipids—inhibits the cell's ability to function properly. This has an impact on all of our organs, including our skin (hence the demand for all those miracle-promising creams and lotions).

So, how does the BioDiet slow and perhaps even reverse some of the effects of aging? It helps prevent mitochondrial dysfunction, thereby limiting cell damage. It reduces the rate of AGE production, by controlling blood glucose. And by spurring the production of ketones, it increases the activity of telomerase in our cells (see page 93, "The Chomosomal Theory of Aging"). Now, that's a real fountain of youth.

While I can't promise that the BioDiet will return you to your teenage years (and really, who wants to go back there anyway?), I can say that during the next 12 weeks on the BioDiet, you will:

- lose weight;
- increase your vitality;
- improve your brain health;
- reduce your risk of developing cardiovascular disease, diabetes, dementia, and perhaps even cancer; and
- yes, look and feel younger than you have in years.

As we conclude part 1, I want you to take a minute to think about all you've learned since you started reading this book, and to appreciate how much better informed you are than when you began. You've come to understand the complex link between the traditional high-carbohydrate diet most of us have been eating all our lives and the chronic diseases that plague us. You know a lot more about how your body processes the food you eat, and why a high-fat, low-carbohydrate diet can be a much better option not only in terms of weight but also overall health. It was a lot to take in, but now you're ready to incorporate this new way of eating into your life. In the coming pages, you'll put into practice everything you've learned thus far, and in doing so, set yourself on the path to optimal health and well-being.

2

How—
Five Steps
to Success

5

Step I: Bio-Assessment
Starting Your Journey

The first step toward getting somewhere is to decide you are not going to stay where you are.
ANONYMOUS

IF YOU'VE SKIPPED straight past part 1 to this "how-to" section, I understand. You picked up this book because you're eager to make a positive change, and you want to do it ASAP. That kind of enthusiasm is great! I wouldn't have written this book if I didn't strongly believe that the BioDiet is your best path to sustained weight loss and improved health. Since most journeys are made easier with a good plan, I've developed five steps to help you through the BioDiet process: Bio-Assessment, Bio-Preparation, Bio-Adaptation, Bio-Rejuvenation, and Bio-Continuation.

This first step, Bio-Assessment, is simple but important; it provides you with a valuable picture of your present state of health and helps determine if the BioDiet is a good fit for you. All you need is a measuring tape, a notebook, a calculator, and a desire to change your life for the better.

Is the BioDiet Right for You?

Although the vast majority of people—about seven out of eight, in my experience—will benefit from the BioDiet. Some will not lose weight, others might not experience any significant health benefits, and a small minority should not be on the BioDiet at all. It's vital that you discuss your plans to adopt the BioDiet—or any ketogenic diet—with your physician. If you are presently being treated for any chronic disease or are recovering from an acute condition, it is of particular importance. But even if you are of normal weight, in good health, and not taking any medications, it is possible that you have an underlying condition that would make you a poor candidate for a ketogenic diet, and only your physician can help you to determine that.

Contraindications

Let's start with the small minority who absolutely should not adopt the BioDiet. There are some conditions, mostly metabolic disorders, that make it impossible for people to metabolize ketones and other by-products related to these metabolic changes. One possible result, for example, is a condition called ketoacidosis, in which dangerously high levels of ketones in the blood can lead to coma and death (see page 74). The following list of contraindications is a modified version of one that first appeared in the journal *Epilepsia* in 2009 by Dr. Eric Kossoff of Johns Hopkins University, an authority on the therapeutic use of ketogenic diets.

- carnitine deficiency (primary)
- carnitine palmitoyltransferase (CPT I or II) deficiency
- carnitine-acylcarnitine translocase deficiency
- beta-oxidation defects
- short-chain, medium-chain, or long-chain acyl-CoA dehydrogenase deficiency

- medium-chain or long-chain 3-hydroxyacyl-CoA dehydrogenase deficiency
- pyruvate carboxylase deficiency
- porphyria
- carriers of the ApoE4 gene
- those taking SGLT2 inhibitors

This is not intended to be a complete list. Consult your doctor to discuss whether or not the BioDiet is right for you.

Requiring Physician Approval

There are also those for whom the BioDiet might not be advised or be of benefit. If you suffer from any of the following conditions, you require assessment, consideration, and oversight by a physician, and it would also be a good idea to seek the advice of a nutritional counselor who specializes in ketogenic diets.

- kidney disease or the presence of kidney stones
- gastrointestinal disorders
- pancreatic disease
- liver or gallbladder disease
- dyslipidemia
- difficulty swallowing
- cardiomyopathy
- chronic metabolic acidosis
- low BMI or a history of eating disorders

Complicating Factors

Those in this group might benefit from the BioDiet, but you should consult with your physician before embarking on it if you have had or currently have any of the following:

- diabetes (type 1 and gestational)
- hypertension (you could experience a significant drop in blood pressure)
- muscular dystrophy
- gastric bypass surgery
- abdominal tumors
- impaired gastrointestinal mobility
- pregnant or breastfeeding

Considering the aging demographic and prevalence of chronic disease in today's population, many of you will have one or more of these conditions. The good news is that the BioDiet might be able to help.

Perform a Self-Assessment

In addition to speaking with your physician prior to embarking on the BioDiet, it's important to do a self-assessment to provide a valuable baseline that prepares you for the changes to come. There are five very simple measurements that aid in determining your body's present state—what is called your morphology, consisting of your size, shape, and fat/lean mass.

1. weight and height
2. appearance
3. waist-to-hip ratio
4. body mass index (BMI)
5. relative fat mass estimation

Weight and Height

Let's start with the easiest evaluation: taking careful note of your physical condition. (I said "easy," not "pleasant"!) I'd suggest logging this information in a notebook, or digitally, where you can refer back to it and add further notes as you progress through the BioDiet steps.

Depending on the time of day you weigh yourself and the state you are in (dressed or undressed, full or empty stomach, pre- or post-workout), there can be significant fluctuations in the number you see on your scale. In order to get the most useful and consistent measurement, get on the scale naked, first thing in the morning, after you've urinated but before you've eaten or taken in any fluids. Note your weight, and then put it out of your head for the time being. It doesn't help to obsess about that particular number.

To measure your height, you'll either need a fancy scale like the one in a physician's office, or you can use the old line-on-the-doorframe technique, something best done with the help of a buddy. Stand barefoot on a hard, flat surface with your back against the frame or wall and your feet together. Get your partner to draw a line on the wall at the top of your head. Then, using either inches or centimeters, measure the distance to the floor and jot that number down alongside your weight.

Appearance

Next up is an assessment of your appearance. Stand naked in front of a full-length mirror and take a good, long look at yourself. If you're in a space with multiple mirrors, like a dressing room in a clothing store, that's even better. (Or perhaps worse!) Check out your entire body from all angles and try not to be judgmental, especially if you are older and gravity has taken its toll. Instead, look at your overall shape. Is it obvious that you're carrying more fat than you should? Where is it accumulating? Pay particular attention to your midsection. Is it tire-like and flabby? Or rounded and firm? Or is most of your weight on your hips and buttocks? Write down what you see.

Waist-to-Hip Ratio
Now that you've noted your weight, height, and appearance, it's time to measure your torso. With your shirt or top up, wrap a measuring tape around your waist at its broadest point, usually around the navel, making sure the tape remains parallel with the floor. Next,

I Feel Your Pain

If you're cursing my name as you work your way through these assessments, it may help to know that I've been there, done that. Once, in my pre-BioDiet days, I was staying at a hotel with a nice pool and posh spa. Long story short, I ended up in the changeroom in the buff and surrounded by mirrors. That's where I experienced my moment of truth.

From the front, things were more or less okay. I was a little thick around the midsection—that "love handles" look. But hey, I was in my 40s. I could live with that. It was the side view that sent me reeling. I could see my upper-gut flab and an enlarged belly, and I even had the beginnings of man-boobs. The rear view was no better, what with the back fat and all. (I know I said not to be judgmental, but I also know that's hard sometimes.)

Until that cringe-inducing moment, I'd considered myself to be a healthy guy. I ate well (or so I thought), I exercised, and I saw my physician for regular checkups. But the naked truth was that I was on the road to being overweight, which I knew could lead to obesity and disease. It was time for a U-turn.

measure your hips at their widest point. Women tend to carry more fat around their hips and buttocks, whereas men tend to accumulate fat around their bellies. The latter correlates more strongly with poor health outcomes; sorry, guys.

To determine your waist-to-hip ratio, divide your waist measurement by your hip measurement. According to the World Health Organization, a healthy waist-to-hip ratio is:

- 0.90 or less for men,
- 0.85 or less for women.

For both men and women, a ratio of 1.0 or higher is linked to an increased risk of cardiovascular disease, among other health conditions. And it turns out that it's a pretty darn reliable indicator. However, it's also worth noting that a ratio that falls in the low-risk range doesn't mean you're completely out of the woods. Some people are what we call TOFI—thin on the outside, fat on the inside. While they might be less likely to have a heart attack, even these thin people can develop fat-related chronic diseases.

Body Mass Index (BMI)

If you've done the above assessments and you're still not completely sure if you need the BioDiet to help you lose weight, a quick check of your body mass index, or BMI, can help to clarify things. Sometimes we think we're in reasonable shape, but the truth is we're anything but.

BMI, which was devised by Belgian mathematician Adolphe Quetelet in the mid-1800s, estimates our body composition; that is, the amount of fat mass we're carrying compared to our lean mass. Obviously, fat mass refers to how much fat you have—but we're not just talking about the subcutaneous fat that accumulates beneath the skin in areas like the butt, midsection, thighs... name

Some people are what
we call TOFI—thin on the
outside, fat on the inside.
While they might be less
likely to have a heart attack,
even these thin people
can develop fat-related
chronic diseases.

your trouble area. We also need to account for the visceral fat that accumulates around the organs inside the abdominal cavity. As we learned in chapter 2, this visceral fat is of particular concern because it is highly inflammatory and closely linked to cardiovascular disease, type 2 diabetes, and breast cancer in women.

Lean mass is everything other than fat: internal organs, bones, muscle, and the outer layers of your skin. For most of us, muscle makes up a lot of our lean body mass: about 36 percent in women and 42 percent in men. That number can get quite a bit higher, especially if you're particularly buff.

To calculate your BMI, use your weight and height in the following formula:

- Imperial (units are pounds and inches): (weight ÷ height²) × 703
- Metric (units are kilograms and meters): weight ÷ height²

So, say you weigh 165 pounds and are five-foot-four, or 64 inches tall. That's:

$$(165 \div 64^2) \times 703 = 28.3$$

What does that result mean? According to the weight classification categories established by the World Health Organization, a BMI of 28.3 means you are considered overweight. Note that for children and teens, BMI is expressed as a percentile relative to other children of the same sex and age to take into account the weight and height changes during development.

Table 5.1: Body mass index (BMI) classifications for adults (20 years and older).

BMI	Classification
Less than 18.5	underweight
18.5–24.9	normal (healthy) weight
25.0–29.9	overweight
30.0–34.9	obese
35.0–39.9	severely obese
40.0 and over	morbidly obese

Are You Insulin Resistant?

If you are overweight or obese, you might also be insulin resistant, or on the road to being insulin resistant. Even people who are of a healthy weight can be insulin resistant. In chapter 3, we learned that insulin resistance occurs when the cells in your liver, muscles, and adipose tissue no longer effectively respond to the signal sent out by the hormone insulin to uptake glucose from the bloodstream. This causes the pancreas to work over-time to produce enough insulin so that the cells pay attention. When the pancreas can no longer keep up with the demand for insulin and/or your cells respond poorly to the insulin signal, blood-sugar levels gradually rise. While age and genetics play a role in insulin resistance, the condition is largely brought about by excess body weight and a sedentary lifestyle.

Although insulin resistance can only be clinically diagnosed through blood tests, there can be warning signs. They include:

- recent weight gain, especially around the midsection;
- mid-abdominal obesity (i.e., your waist is larger than your hips);

Relative Fat Mass Estimation

Although BMI is used by physicians as a guideline for determining normal or abnormal weight, it's not very reliable when it comes to estimating body fat (or adiposity) as a percentage of total body mass. As a result, it tends to underestimate adiposity in those with less lean muscle mass, especially the aged, and overestimate it in lean, muscular individuals, especially athletes. For example, most football running backs would be considered overweight or obese according to BMI despite being very lean.

A simple and more accurate way to determine body fat is to use a formula recently devised by Orison Woolcott and Richard Bergman at Cedars-Sinai Medical Center. Rather than estimate relative body

- periods of light-headedness or dizziness for no apparent reason;
- shifts in mood, particularly after eating;
- brain fog, poor concentration, and/or poor memory;
- cravings for sweets and breads;
- pronounced mid-morning and mid-afternoon hunger;
- worsening aches and pains;
- growing fatigue;
- increased susceptibility to infectious diseases like colds and flu;
- pimples, skin tags, and dry, irritated skin; and
- dark patches on the groin, armpits, or back of the neck.

Some might say these are normal indicators of aging and we can afford to ignore them. But when you consider that 50 percent of people with insulin resistance will become diabetic within 5 to 10 years, I think it's critical to heed these warnings *before* finding yourself on the road to serious health problems.

mass, this calculation approximates relative fat mass, or RFM. The formula is slightly different for women and men because women have higher amounts of subcutaneous fat, which does not correlate with poor health outcomes.

- Women: 76 − (20 × (height ÷ waist circumference)) = RFM
- Men: 64 − (20 × (height ÷ waist circumference)) = RFM

Let's plug in the numbers (measured in inches) for the average American female and male adult, according to the US Centers for Disease Control and Prevention:

- Women: 76 − (20 × (63.7 ÷ 38.1)) = 76 − (20 × 1.67) = RFM of 42.6%
- Men: 64 − (20 × (69 ÷ 40)) = 64 − (20 × 1.73) = RFM of 29.5%

50 percent of people with insulin resistance will become diabetic within 5 to 10 years.

For the average American female, an RFM of 42.6% means they are considered obese at any age. For the average American male, an RFM of 29.5% means they are either overfat or obese, depending on their age (see table 5.2).

Table 5.2: Body fat ranges for females and males. (*Source: National Institutes of Health and World Health Organization guidelines.*)

BODY FAT RANGES FOR STANDARD ADULTS				
FEMALE				
AGE	**BODY FAT PERCENTAGE**			
	Underfat	Healthy	Overfat	Obese
12–20	<18%	18–28%	28–33%	>33%
21–42	<20%	20–30%	30–35%	>35%
43–65	<21%	21–31%	31–36%	>36%
66–100	<22%	22–32%	32–37%	>37%

BODY FAT RANGES FOR STANDARD ADULTS				
MALE				
AGE	**BODY FAT PERCENTAGE**			
	Underfat	Healthy	Overfat	Obese
12–20	<15%	15–21%	21–26%	>26%
21–42	<17%	17–23%	23–28%	>28%
43–65	<18%	18–24%	24–29%	>29%
66–100	<19%	19–25%	25–30%	>30%

Go ahead and calculate your own RFM and see where you fit. If you fall into the overfat or obese categories, the BioDiet can help you lose that weight and improve your health in the process.

Bring in the Professionals

Did your self-assessment reveal that you could benefit from the BioDiet? That's great information to have, and you're already better informed about your health than you were when you picked up this book. But, as I mentioned at the start of this chapter, it's vital that you bring in the professionals before you dive in to the BioDiet or any other ketogenic diet. Your physician can help you arrange a series of tests that will give you more detailed insight into your present state of health and provide the best benchmarks for evaluating it going forward.

Depending on where you live and the nature of your health insurer, some of the costs for these tests will be covered; others you might have to pay for out of pocket. Note that only a qualified professional—your physician—should interpret these results.

Standard Tests

- Complete blood count: examines the number and health of your red blood cells, white blood cells, and platelets (blood-clotting factors).

- Blood lipids (a.k.a. lipid panel): measures your total cholesterol, low- and high-density cholesterol (LDL and HDL), and triglycerides. If your LDL is high, you may want to consider, if available, a more advanced test that measures particle number and size.

- Blood glucose

  - Fasting glucose: examines how your liver and other organs are releasing glucose into the blood outside of meals. (You'll need to fast for at least eight hours before the test.) If your fasting glucose is high, your physician might follow up with a two-hour glucose tolerance test for a more accurate, short-term measure.

  - Hemoglobin A1c: measures the glucose that clings to the hemoglobin in red blood cells. Because these cells live for about four months, this test provides an estimate of blood-glucose levels over a much longer time period.

- Basic metabolic panel: measures electrolytes—sodium, potassium, calcium, chloride, and bicarbonate—as well as creatinine, albumin, and blood urea nitrogen (note: glucose is also part of the basic metabolic panel).

- Standard liver panel: determines liver function by measuring liver enzymes in the blood including Gamma GT, ALT, AST, ALP, and LD.

- Kidney function and urinalysis

  - GFR (glomerular filtration rate): a standard measure of kidney function.

  - Urinalysis: helps to determine kidney and overall metabolic function. It will also determine if you have excess protein, glucose, hemoglobin, or white cells in your urine, all of which could indicate infection or chronic disease. This is also an opportunity to measure your urine ketone levels, which should be low to start but will rise significantly on the BioDiet.

Additional Tests

- Blood ketones: provides a more precise measurement of your ketone levels and rate of production, which can help determine your degree of adaptation to the BioDiet.

- Magnesium: measures blood level of magnesium, which is useful should you experience muscle twitching or cramping on the BioDiet and require supplements.

- Vitamin D: measures the blood level of this important vitamin for calcium absorption and metabolism. Vitamin D correlates with hormones that affect, in particular, bone density.

- High-sensitivity C-reactive protein (hs-CRP): measures systemic inflammation. Note that high levels could also be a result of an acute infection, so your present health status is important to consider.

- Blood LPS: measures lipopolysaccharide, which is not normally found in the blood, but if present can indicate an inflammatory intestinal condition called leaky gut syndrome. High blood LPS levels have also been suggested as possible aggravating agents in some neurological disorders.

- DEXA (dual-energy x-ray absorptiometry): a bone density test that not only provides a good estimate of the health of your skeleton, something particularly important for post-menopausal women, but is also the gold standard for measuring actual fat mass, so you get two results for the price of one.

Get Your Head in the Game

Self-assessment? Check. Doctor's visit and tests? Check. Nice work! Now that you're physically good to go, it's time to make sure you're also mentally prepared for what's ahead.

If you've worked your way through the first part of this book, it might be easy to conclude that the BioDiet is all about *what* you eat. To some extent that's true: a diet high in carbohydrates poses significant challenges to your health and well-being. But the real power of the BioDiet is in changing *how* you think about what you eat. It requires you to be truly mindful of your food, including where to shop, what to buy, and how to prepare it. At first, it can seem like a lot of mental work, but overcoming these early challenges is integral to your success in losing weight and improving your health.

The real power of the BioDiet is in changing *how* you think about what you eat.

To help you through this period I'm going to suggest two simple words that became a bit of a mantra for me when I embarked on a ketogenic diet: planning and commitment. *Planning* what you're going to eat for the week, or even over just a few days, not only ensures that you have a dietary strategy, it also guarantees that you won't go hungry. And *committing* to the BioDiet for the 12 weeks needed to make a real change to your health means having faith in yourself that you can do this and, perhaps more importantly, that you're worth it. I discuss this in more detail later, but for now just know that I've got your back here.

Social support can be crucial when it comes to making any major lifestyle change. If possible, have someone join you on the BioDiet journey. Enlisting a buddy provides the opportunity to compare notes and gives both of you a shoulder to lean on and a sympathetic ear. And please take advantage of the BioDiet website (BioDiet.org), where you'll find a wide range of articles that answer your questions and keep you on track.

Choose the date you embark on Step II of the BioDiet wisely. Cutting out carbohydrates for the first time is probably not something you want to do during the holiday season, or right before you take that cruise, but January could be a good time to start. If you have a wedding or reunion in the future, you might want to count back 12 weeks from that date and begin the BioDiet then. Nothing like arriving at a special event looking and feeling like a million bucks.

Keep in mind that while Step II—which lasts two weeks and involves a gradual transition—shouldn't be too difficult, Step III can be more onerous. It also lasts about two weeks, but because it involves the short-term elimination of almost all carbohydrates from your diet, I'd suggest clearing your calendar of dinner parties and other temptations. By the time you get to Step IV, you'll be on your way to a slimmer, healthier, more energetic you. Step V is all about maintaining the BioDiet so you continue to reap its benefits.

Finally, let's introduce some incentives. Like all animals, we humans are motivated by reward. So what will yours be at the end of each step of the BioDiet? What's the one thing that will motivate you to stick with it even on those days when going out for a burger and a beer seems a lot more attractive than making yourself a healthy low-carb meal? Make the reward for completing Step II small but meaningful. Consider a reward that involves an experience rather than stuff—going to a sporting event, for example, or taking in a movie (without the popcorn, of course). For Step III, consider something a little more substantial, like a concert or a day at the beach or on the slopes. The completion of Step IV is, in my books, cause for real celebration. Take a trip, have a party, splurge on a fancy new fill-in-the-blank. (I was hoping to reward myself with a Lamborghini at the end of the 12 weeks—so much for preferring experiences—but figured it was more prudent to go out for a nice dinner.)

While you give some thought to what your rewards will be, also take a moment to give yourself credit for assessing your health and putting into motion a plan to improve it. Sometimes the first step is the toughest one to take.

Next up is Bio-Preparation: the important groundwork for the significant changes ahead.

6

Step II: Bio-Preparation
Getting Ready for Takeoff

*Before everything else, getting
ready is the secret of success.*
HENRY FORD, AMERICAN INDUSTRIALIST,
1863–1947

I LEARNED THE IMPORTANCE of being prepared early in my life from my father, Mirv. He was an airline pilot, and before every flight he'd draw up a plan based on the type of plane he'd be flying, its departure and arrival times, the amount of fuel needed to get to the destination (running out of gas midair is never a good idea), and any problems that might be encountered along the way, like wind and weather. The plan also indicated the route the plane would take; should they find themselves off course, they had the information they needed to get back on track.

Step II of the BioDiet is your flight plan. It's a two-week preparation period that introduces you to a new way of eating, makes sure you have enough of the right fuel to get you where you're going, helps you navigate any turbulence you might encounter,

The more you know
about how the BioDiet
works, the more likely
you are to succeed
at losing weight and
improving your health.

and establishes a logbook to note your progress. This step also prepares you for the significant nutritional changes and for the lifestyle
changes that help ensure your continued success in losing weight
and improving your health. I've included this important step for a
few reasons. By easing your way into the BioDiet, you increase your
chances of sticking with it and decrease the likelihood of experiencing unpleasant side effects. You also have the opportunity to get
ready for Step III (Bio-Adaptation) by developing meal plans, preparing your pantry, adjusting your social schedule, and generally
getting your head in the game.

Everyone approaches change in their own way. Some like to
dive right in; others wade in a bit more slowly; and still others take
considerable time before even dipping a toe in. For the reasons
mentioned above, I strongly recommend a "go slow" approach,

which is why I've developed a two-week Bio-Preparation step. That said, I know that some of you will want to move faster. If you really feel that you can't wait, you can try an intermediate one-week preparation period, but please do so knowing that it can be quite a bit more challenging. (Some of the changes you're about to make can be uncomfortable.) Finally, if you're dead set on forgoing Step II altogether, be forewarned that diving right in can be a shock to the system. Trust me on this one. When I first adopted the BioDiet, I dove straight in at Step III; though I surfaced on the other side, it wasn't a lot of fun. I share some of those experiences in the next chapter. In the meantime, let's start with a practical look at what to expect in the days ahead.

Bio-Preparation: An Overview

Making significant changes to the way you eat can seem overwhelming—and make no mistake, the changes you make in this stage are, indeed, significant. By the time you complete Bio-Preparation, you will have eliminated sugar, cut back on alcohol and many starchy foods, and started the process of replacing those unhealthy calories with healthy fats. In an effort to take a bit of the mystery out of the process, here are answers to some of the most common questions I'm asked about this important step.

What changes will I be making to my diet?
Not many in week 1. We start by removing the major dietary culprit: sugar. We also cut back on alcohol and increase your water intake. And we introduce your body to probiotics and your cells to their preferred fuel: ketones. Week 2 involves replacing the empty calories of high-glycemic index starchy foods with healthy fats. We also start to fine-tune when you eat, which helps you develop healthy habits.

What do I need to buy?

During week 1, you need MCT (medium-chain triglyceride) oil or another ketone supplement. I also recommend probiotics and a multivitamin. You need to shop for the food you'll be eating in Steps II and III (see the grocery list in appendix A), and, since you'll be eliminating sugar, you'll want to pick up a sugar substitute (see the list of sweeteners in appendix B). I recommend you try a few of them until you find one you like.

What else do I need to do?

If you haven't read part 1 of this book yet, that's okay, but now is the time. I strongly believe that the more you know about how the Bio-Diet works, the more likely you are to succeed at losing weight and improving your health. This is also the time to see your physician, if you haven't already done so.

Now is also the time to choose a system to keep track of things. A calendar (electronic or otherwise) is helpful for meal planning and preparation, and I suggest using your notebook from Step I to write down any positive or negative side effects you may experience, as well as any challenges or successes. On day 1, weigh yourself on a reliable scale first thing in the morning, after urinating. Record your weight and body measurements in your diary, as described in Step I. This information is important in charting your progress.

If you already exercise, keep it up. If you don't yet have an exercise regimen, you may want to begin by introducing some moderate exercise like walking, bicycling, or swimming into your schedule.

As part of Step II, I ask you to rid your home of any foods that won't be in your diet during the following weeks. Some people prefer to use them up during this Bio-Preparation period, but do yourself a favor and don't double up on your cookie consumption in an effort to get them out of your pantry. Not only will this increase your weight, but it will also make your transition to Step III more difficult.

What physical and psychological changes can I expect?
Unless you already stay very well hydrated, the increase in water consumption during this stage will require more frequent urination. Increased water intake also tends to promote bowel movements, and the introduction of MCT oil (which I talk about later) can make the need to find a washroom even more urgent. Plan your day, especially any travel, accordingly. The probiotics you introduce should alleviate some of the intestinal urgency, but they can take several weeks to work.

If you currently consume a lot of sweet drinks and foods, you will almost surely experience some sugar cravings during this step. Even those who don't have a sweet tooth may find this to be the case, given that sugar is added to even savory processed foods. You might experience headaches as a result of eliminating sugar, but once your body has adjusted, these should disappear.

The longer you are on the BioDiet, the easier it will become to resist sugar and other high-carbohydrate foods.

Prolonged carbohydrate cravings could cause you to question whether you want to continue on the BioDiet. Recognize that this is your brain's way of convincing the rest of your body to provide what it wants, which is more glucose. Don't be hard on yourself. Just do your best not to give in.

One of the goals of the BioDiet is to make us more mindful about what we eat. This is especially important given that fully half of our meals are consumed outside of the home, and half of those are eaten at fast-food restaurants. The opportunity to prepare food at home gives us a better appreciation for what we eat and offers valuable insights into our food habits.

Bio-Preparation Basics: Laying the Groundwork

Bio-Preparation is, as the name suggests, all about laying the groundwork for this new and healthier way of life. An important part of this step is eliminating those foods you know aren't good for you, but you just can't resist. Before I started the BioDiet, I would regularly veg out on the couch and eat half a bag of Italian biscotti. I knew it was unhealthy, but, hey, they were in the cupboard just waiting to be eaten! And once you eat one... Whatever your kryptonite, these two weeks are when you get it out of the house. I suggest you tackle this task straight off the bat. Not only will it free up refrigerator and pantry space for all the good food you'll be eating over the next 12 weeks, it'll also ensure you don't fall off the wagon just as you've climbed on board.

Kitchen Detox

- The first step in your preparation? Rid your home of all forms of sugar, including brown and white sugar, honey, molasses, maple syrup, and corn syrup. Toss out any candies, cookies, cake, or ice cream. Ditch the sweetened cereals as well as granola and energy bars, which, despite their healthy connotation, are anything but. Pour out any sweet beverages, including flavored coffee, bottled iced tea, chocolate milk, and sodas. And finally get rid of fruit juices, which we know are packed with carbohydrates in the form of fruit sugars. (If the idea of ditching all of this food is making you cringe, consider giving it away.) As you're riffling through your cupboard, check nutrition labels to determine the sugar content of products. (Find a guide to reading labels in appendix C.) When looking at a product's ingredients list, keep in mind that food manufacturers cleverly use a variety of different sweeteners, from agave nectar to golden syrup. Don't be fooled: these are sugar by another name. (You'll find a list of the 60-plus sugar aliases at BioDiet.org.)

- If you have kids at home or other family members who aren't joining you on the BioDiet, you may not be able to ditch all these items without becoming extremely unpopular. If that's the case, move them to a cupboard or refrigerator shelf that you now consider off-limits. Clearly explain to the household why you're eliminating this food from your diet and extract a promise from everyone that they won't try to tempt you into eating it. (Kids will be kids, after all.) The longer you are on the BioDiet, the easier it will become to resist sugar and other high-carbohydrate foods.

Now is also a good time to be on the lookout for foods to toss in week 2—starches, which can be broken down into four categories:

1. grains, including rice (and rice cakes), barley, oats, and quinoa;
2. starchy vegetables including potatoes, sweet potatoes, yams, parsnips, turnips, beets, carrots, winter squash, and corn;
3. foods made with flour, including bread, pasta, and crackers; and
4. legumes such as beans, peas, and lentils.

Make a mental note of these items, and be prepared to part with them sometime during week 2. Of course, if you've opted for the one-week approach to Bio-Preparation, you should go ahead and clear these out at the same time as your sweets.

Kitchen Restock

Talk about a fresh start! Now that you've rid your kitchen of sugars, your refrigerator and cupboards may be looking a bit bare. Time to hit the grocery store, where you can stock up on plenty of much healthier options. You've heard the old tip about shopping the perimeter of the store, where non-processed foods tend to live? Well, it's good advice. For a full grocery list, please see appendix A, but here's a quick rundown of what can be added to your cart.

- Plenty of vegetables loaded with vitamins, minerals, fiber, and healthy phytochemicals. These include artichokes, asparagus, avocado, bell peppers, bok choy, broccoli, brussels sprouts, cabbage, cauliflower, cucumber, eggplant, green and yellow beans, kale, lettuces, mushrooms, olives, onions, radishes, snow peas, spinach, summer squash, tomatoes, watercress, and zucchini.

- High-fat dairy products including cream, sour cream, cream cheese, yogurt, kefir, and cheeses like cheddar, gouda, Swiss, Parmesan, Brie or Camembert, blue, Jarlsberg, goat cheese, feta, and mozzarella.

- Proteins such as meats, poultry, eggs, tofu, tempeh, textured vegetable protein (TVP), fish, and shellfish. Canned tuna, salmon, sardines, and crab are also very useful to have on hand.

- Fats including olive, macadamia, avocado, and coconut oils; grass-fed butter, and ghee (clarified butter).

- Nuts such as pecans, macadamia and Brazil nuts, hazelnuts, and walnuts.

- Raw seeds including pumpkin, sunflower, and sesame seeds.

- Unsweetened almond butter.

- An orange, kiwi, or apple, and small amounts of raspberries, blackberries, strawberries, and blueberries.

- Sugar-free candy and/or chocolate (in the diabetic section of your grocery store or pharmacy). Since some sugar substitutes can cause gastrointestinal issues in some people, I suggest you experiment with a few and choose one that works best for you.

While you're out on your healthy shopping spree, you can also pick up these supplies at the pharmacy (see "Supplements" for more on why they're recommended): MCT oil, probiotics, fish oil, and multivitamins.

Supplements

I'm a great believer in getting all the vitamins and nutrients you need from food. The BioDiet is an excellent way to do that. However, your body takes a little time adjusting to a diet high in fat and low in carbohydrates. Here are a few things that can help with that process and also benefit you in the long term.

MCT oil: I recommend you introduce MCT (medium-chain triglyceride) oil into your diet during Steps II and III. MCT oil goes straight to the liver, where it is either burned or converted directly into ketones. It's an excellent way to get your body accustomed to the elevated level of ketones that is one of the many benefits of the BioDiet. I prefer an oil that contains 100 percent caprylic acid (an 8-carbon, medium-chain fatty acid), but one that has a mix of MCTs is also fine. For some people, MCT oil can result in a sudden urge to visit the bathroom, so start with one teaspoon first thing in the morning and work your way up to as much as one tablespoon. Increase it gradually, though, especially if you have a long commute to work.

Probiotics: If you do experience intestinal problems, you may want to consider taking a probiotic, which can help establish and maintain a healthy gut. Look for *Bifidobacterium infantis* on the label or list of active ingredients. Note that it can take three or four weeks for the good bacteria to begin to work. If you have a pre-existing illness, are recovering from surgery, or are immune compromised, check with your physician to be sure probiotics are safe for you to use.

Fish oil: This is a great source of essential omega-3 fatty acids that have unpronounceable names—eicosapentaenoic acid and docosahexaenoic acid—and so are usually referred to as EPA and DHA. Both have been shown to reduce inflammation and so contribute to healthy aging and reduced incidence of chronic disease, especially cardiovascular disease. Recent research has also provided promising results for weight loss and the prevention of Alzheimer's disease. I recommend taking at least 250 mg/day each of EPA and DHA, preferably more.

Multivitamins: Although the BioDiet is a very nutritious way of eating, I suggest you take an iron-free multivitamin in the morning to be certain you are getting all the vitamins and minerals you need. Look for one that contains calcium and magnesium, or take a separate cal-mag supplement that also contains vitamin D, which aids in calcium absorption. Vitamin D may also help to lower blood pressure and could reduce your risk of heart attack, cancer, and arthritis. (I talk more about the importance of both dietary minerals and salts in chapter 7.)

Week 1: Saying "So Long" to Sugar

Regardless of whether you've chosen the one- or two-week approach to Bio-Preparation, let's get to it. You already know that this first week is all about removing sugar from your diet. (If you need a refresher as to why eliminating carbohydrates, especially sugar, is important, revisit chapter 3, or check out "The Truth about Fructose" on page 162.) But there's more to week 1 than kicking your sugar habit. This is also a time to create new habits and to prepare for the bigger changes coming in Step III. You've already started this process by detoxing and restocking your kitchen. Now you're ready to put all of that preparation into action.

This is the point where some people do a double take. The stuff in your refrigerator looks different. Your favorite recipes involve pasta, or rice, or bread. Your go-to coffee-break snack is a vanilla latte and a muffin. The more you think about the changes you're about to undertake, the scarier they seem. *How do I do this?* you ask yourself.

Never fear. I'm going to tell you exactly how to do this. By the time you're finished reading this chapter, you'll feel confident in your ability to conquer this step and subsequent ones, and you'll be well on your way to experiencing all of the health benefits that come as a result.

Managing Your Day to Maximize Weight Loss

One of the keys to success on the BioDiet is learning how to manage your day to your advantage. *When* you eat or drink is, in many ways, as important as *what* you consume. Keeping a picture of an annotated 24-hour clock (see figure 6.1) taped to the refrigerator, or tacked into your journal, is a helpful way to remind yourself of the basics.

Figure 6.1: Optimal times for your daily BioDiet routine.

Manage your day to maximize weight loss

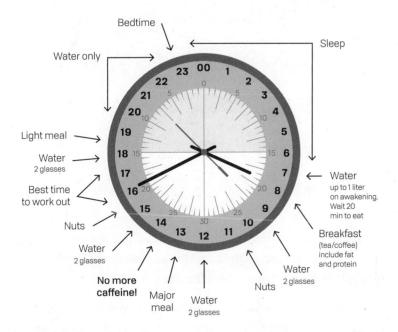

- Drinking water first thing in the morning hydrates you after an extended period without fluids. And by expanding your stomach, water also creates a full feeling, which helps prevent you from overeating at breakfast.

- The morning is a great time to introduce protein and fats, nutrients that will give you energy and keep you fueled until lunchtime.

- Ensuring you get enough salt is also a crucial part of the Bio-Diet. Salted nuts mid-morning or afternoon are a great way to boost not only your salt intake but also to ensure you are getting enough healthy fats.

When you eat or drink is, in many ways, as important as *what* you consume.

- Eating your main meal at lunch, so long as it's a low-carbohydrate meal, helps you avoid the usual afternoon slump, and by eating a lighter meal at supper, you sleep better.

- According to the American Alliance for Healthy Sleep, caffeine ingested six hours before bedtime can reduce total sleep time by an hour, so go easy on it, especially late in the day. As a general rule, don't drink anything caffeinated after 2 p.m.

- The benefits of afternoon workouts include an increase in overall performance and, because you're already warmed up by your day's activities, a decreased chance of injury. That said, the most important consideration when it comes to exercise is that you do it! So find a time that works best for you.

- A big supper or snacking late at night can raise your blood-sugar levels, slow your metabolism, and result in GERD (gastroesophageal reflux disease, also known as acid reflux). Opt for a lighter, smaller meal and eat it early to give your body time to digest the food before you turn in for the night.

- Drink enough water before bed to keep you hydrated but not so much that you're up repeatedly in the middle of the night to use the bathroom. Go to bed at a decent hour and, hopefully, sleep like a lamb.

With these basics in mind, let's take a closer look at how a day's meals might look during week 1.

Breakfast

When discussing meals with those I've counseled, one of the most common questions people ask is "What do I eat for breakfast?" Breakfast, it seems, is a big challenge. Maybe that's because, for many of us, a typical morning meal consists largely of carbohydrates such as cereal, toast, muffins, or breakfast sandwiches. If you opt for seemingly healthier options like sweetened yogurt, fruit, oatmeal, or granola, you might be getting more nutrients, but you're also getting a ton of extra sugar—and that's what we're trying to cut back on in week 1.

So, if sugar's out, what's in? Here are some suggestions. (If you're reducing the Bio-Preparation period to one week you'll want to start removing cereal and bread from your diet now.)

- Unsweetened cereals including steel-cut oatmeal with added fruit and full-fat milk or cream.
- Whole-grain toast or bran muffins with unsweetened almond butter.
- Celery sticks with full-fat cream cheese or almond butter.
- Full-fat yogurt with berries, toasted nuts, and seeds (see "BioDiet Yogurt" on page 134).
- Tomato and avocado with meat, fish, or cheese.
- Lettuce/radicchio roll-ups or endive leaves with tuna salad.
- Pancakes or waffles, with yogurt and berries instead of syrup, and a side of sausage.
- Bacon and eggs with tomato, avocado, and toast with nut butter.

I've put bacon and eggs last for a reason. There's a persistent perception out there that if you're on a ketogenic diet, bacon and eggs are the only things you can eat for breakfast. Clearly that's not the case. There has also been much debate over whether eating eggs every day is bad for your health, because eggs contain cholesterol and are therefore suspected of boosting cholesterol levels in the blood. However, most of the cholesterol in your blood is produced in the liver, and for most people there is no relationship between it and the cholesterol we consume in our food. That said, if you have questions about your cholesterol levels, or any other health concerns, I recommend you discuss them with your physician.

BioDiet Yogurt

When I set out to write *BioDiet*, I knew that I wanted it to be about the science behind ketogenic diets rather than a book of recipes. That said, I'd like to pass on one breakfast tip that you might find helpful—a surprisingly simple way to turn plain yogurt into something really tasty.

Start with unsweetened yogurt that has the highest milk-fat percentage you can find—10 percent or above is ideal. Don't be overly concerned if the nutrition label indicates that the yogurt contains some sugar; this is lactose, almost all of which was converted during the fermentation process to the lactic acid that gives yogurt its tart taste. Stir in your favorite sweetener (you'll find a list of them in appendix B) and some natural vanilla extract, or whatever unsweetened flavoring you prefer. Top it off with some berries and a combination of toasted pumpkin, sunflower, and sesame seeds and some chopped nuts. It's a breakfast worth getting out of bed for.

If you're a person who typically isn't hungry in the morning, that's okay. Be sure, however, to start your day with at least two glasses of water, and to drink water regularly throughout the day. This is especially important on a ketogenic diet because decreased carbohydrate consumption has a diuretic effect, which I discuss in detail in chapter 7.

Mid-Morning

In Step II, you are consuming less sugar than you perhaps have in the past, but your diet is likely still high in carbohydrates. This means that there's a good chance you still experience that all-too-familiar mid-morning hunger and energy slump. This is due, in part, to that fact that we're generally more sensitive to insulin early in the day, which means it has a greater effect, pulling more glucose from the blood and lowering our blood sugar. Replace that mid-morning banana bread with a small handful of nuts. I carry a small bag of them with me wherever I go, and they've saved my bacon more than once. If you're still hungry, a small piece of high-fat cheese is a good option or a little fruit. If you have coffee or tea, replace fat-free, low-fat, or whole milk with cream. Not only does cream have just 3 grams of carbohydrates per cup compared to whole milk's 13 grams, it also contains the fat your body will soon be burning in place of carbohydrates. And don't forget to hydrate, which also helps fill you up.

Lunch

Lunches on a ketogenic diet are generally easier to manage than breakfast, when we're habituated to eating something sweet. If you've adopted a two-week Bio-Preparation, have soup and a sandwich. If you've chosen the one-week approach, cut way back on the bread. In either case, you won't be eating any bread in Step III, so you may want to start replacing sandwiches with salads that include a protein such as eggs, chicken, fish, tofu, or shellfish. Bun-less burgers are also an option, as long as you forgo the fries and order a salad. Since commercial salad dressings often contain sugar, opt for olive oil and vinegar instead.

We all know that soda contains a lot of sugar (about 9 teaspoons per can; that's a whopping 36 grams of carbohydrates!), but other beverages can be almost as bad. Some seemingly healthy options—like organic lemonade or sparkling citrus drinks—can contain as much as 8 teaspoons (32 grams) of sugar per single serving. If you're really craving a soda, make it a diet one, though my hope is to wean you off the need for that super-sweet taste in the weeks to come. Sparkling water with a little lemon or lime is an excellent alternative.

The benefits of even a little bit of exercise are significant, so consider a quick walk before you head back to work. The Italians call this a passeggiata, and it's their secret to good health. The walk aids in digestion and boosts your metabolism, thereby increasing the amount of fuel you burn. It also encourages your body to burn any sugar you've eaten rather than store it as fat.

Mid-Afternoon

Oh, the mid-afternoon munchies. We've all been there, haven't we? Afternoons can be full of temptation: the plate of homemade cookies your well-meaning co-worker brought in to share, or the display of baked goods in full view of the cash register at the coffee shop. This is a good time to practice commitment, because the sooner you forgo sugar, the faster your body will transition to burning fat and the quicker you'll lose weight and get healthy. If you find you must eat something sweet during Bio-Preparation, opt for a small piece of fruit.

The BioDiet and Exercise

Earlier in the book, I mentioned the 80:20 rule that suggests what we eat (or don't eat) is responsible for 80 percent of weight loss, while exercise contributes just 20 percent. Like diet, however, exercise has many benefits over and above the numbers on your bathroom scale. Regular exercise improves mood and energy levels; reduces feelings of anxiety and depression; has a positive impact on memory and brain health; increases muscle mass and bone density; reduces the risk of chronic diseases like diabetes, obesity, and heart disease; helps you sleep better; and even enhances your sex life. It's safe to say that if exercise was bottled for sale, it'd be flying off the shelves!

If you don't already have an exercise regimen, the Bio-Preparation step is a good time to start one. You can begin by checking out your community center or local gym, and by giving some thought to what kind of exercise you'd like to do on a regular basis. I recommend a combination of aerobic exercise (including walking, running, or swimming), strength training (weight lifting and resistance exercises, including suspension training), stability programs (such as Pilates), and flexibility training (including yoga and stretch classes). Before you embark on any of these, speak with your physician, particularly if you have health issues, including obesity, or are over the age of 45.

Both the American College of Sports Medicine and the Canadian Society for Exercise Physiology recommend 150 minutes of exercise per week, including everything from brisk walking to jogging to bicycling to work. Before you get going, establish some realistic goals. Then start slow and increase the frequency and intensity of your workouts as you increase your fitness level. And don't forget to let your body recover between workouts. The more deserved your rest, the better it will feel.

Dinner

Dinner is more or less business as usual but without sweetened drinks, or dessert. If you have opted for a one-week Bio-Preparation period, you should also cut back on bread, potatoes, rice, pasta, and noodles. If it's your routine to enjoy a glass of wine or a beer with your meal, you can still do that, though you should be consuming no more than two drinks a day. For the one-week prep period, limit alcohol consumption to one drink. This helps you prepare for Step III, when you'll be abstaining from alcohol altogether. Remember to drink lots of water.

Most of the cholesterol in your blood is produced in the liver, and for most people there is no relationship between it and the cholesterol we consume in our food.

Post-Dinner

The post-dinner period can be a challenge, especially if you are accustomed to vegging out in front of the TV eating bite-sized brownies (or Italian biscotti). Take another passeggiata and, if you find you're still craving sugar, indulge in a small low- or no-sugar treat. Semi-sweet dark chocolate is a good choice or, better still, have a little of the sugar-free chocolate or candy that's now in your pantry. However, remember that the sweeteners typically used in these products can cause digestive issues in some people, so the safer bet might be a slice or two of apple.

Bedtime

If you refrain from eating anything two hours before you go to bed, you'll sleep better. And believe it or not, sleep is particularly import-ant when it comes to losing weight. That's because decreased sleep or poor sleep quality increases hunger and appetite and decreases glucose tolerance and insulin sensitivity—hence the growing num-ber of studies that link sleep loss with obesity. Have a glass of water at your bedside—and not just to sip it during the night. It will also act as a reminder, the following morning, to drink water before you go about your day.

Taking Stock: Week 1 Wrap-up

Whether you adopted a one- or two-week Bio-Preparation step, now is a good time to give some consideration to how you're feeling. Are you happy about the changes you've made so far to your diet and lifestyle, or are you wondering what you got yourself into? Perhaps you have a foot in both worlds—missing your carbohydrate-rich comfort foods while at the same time feeling good that you're doing something positive for your health. Whatever it is you're feeling, I get it! The BioDiet requires you to make a radical change not only to what you eat but also to the way you think about what you eat. And we humans aren't, for the most part, big fans of change.

It's also important at this point to consider how you feel physically. Are you experiencing headaches or dehydration, or are you having trouble sleeping and feeling a little cranky as a result? These and other minor side effects are completely normal and should resolve themselves fairly quickly, but if you have larger, more serious concerns, be sure to discuss them with your physician.

The end of week 1, particularly if you've opted for a one-week Bio-Preparation step, is a good time to redo the measurements

Nature's Candy

I'm often asked whether it's okay to eat fruit on a ketogenic diet. The answer is yes—and no. Fruits contain sugar, and some contain a lot. A large apple, for example, can have 23 grams of sugar, and the same goes for a cup of mango. Since Step III of the BioDiet is fruit-free, it's important that you reduce your consumption of it here in Step II. If you've chosen a one-week Bio-Preparation step, cut back on fruit even further. As a general rule, choose fruit with the highest nutrient density, greatest amount of fiber (especially soluble fiber), and the least amount of sugar. Good examples include raspberries, blackberries, strawberries, and blueberries. Dried fruit is off-limits; it's very high in sugar. A 1-ounce serving of dried apricots contains 15 grams of carbohydrates, dried pears nearly 18 grams, and raisins 22 grams, 17 of which are sugar! Since manufacturers often add even more sugar to dried fruits (such as blueberries, cranberries, and especially dates), even a small portion can have a big impact on your blood-glucose level.

you compiled back in chapter 5 and make note of them. Don't be surprised if you haven't lost weight or inches yet. That is normal. Bio-Preparation is designed to ease you into the BioDiet, and you're unlikely to see any significant weight loss until you've moved to Step III's Bio-Adaptation. In the meantime, take pride in the positive life-style changes you've accomplished so far. Now, on to week 2.

Week 2: Sticking It to Starch

If week 1 of Bio-Preparation is largely about eliminating sugar from your diet, week 2 is about eliminating starch. You'll recall that starches are polysaccharides ("many sugars") composed of glucose molecules bonded together in long strands that are broken down in our digestive tracts and absorbed into the blood, resulting in an increase in blood sugar. That's why it is surprising—shocking even—that at the time of writing, the United States Department of Agriculture is still advocating filling one-quarter of their recommended "MyPlate" with starch. Canada's new Food Guide is even more out of touch, suggesting that up to three-quarters of our calories come from carbohydrates, the majority of which are starch. Fats, meanwhile, barely rate a mention. The BioDiet is designed to turn those guidelines upside down by all but eliminating starch from our diet and replacing those calories with healthy and very satisfying fats.

If forgoing starchy foods seems a bit rash, consider that in Step III you'll be reducing your carbohydrate intake to 20 grams or less per day (compared to the 300 or more grams many of us now eat). By eliminating sugar, you've gone a long way toward that 20-gram goal, but in order to reach it, you need to take the next step by eliminating the worst of the starchy culprits—foods with high glycemic index, or GI (see "The Glycemic Index," page 146).

Alcohol and Carbohydrates

Although distilled spirits have calories, they don't contain carbohydrates, so if you're a scotch and soda drinker, you're in luck. The dry vodka martini preferred by James Bond (either shaken or stirred) has only 0.4 grams of carbohydrates, from the vermouth. A mixed drink made with soda or juice is an entirely different matter. A gin and tonic packs 16 grams of carbohydrates in every drink; a screwdriver, 18; and a mojito, about 24.

Prior to fermentation, wine and beer contain lots of carbohydrates from the grapes and grains. During fermentation, however, the yeast metabolizes the sugars, producing alcohol (along with heat and carbon dioxide). The residual sugars are what contribute to the beverage's total amount of carbohydrates. Amber ales and honey lagers can contain as much as 17 grams of carbohydrates in a 12-ounce serving, whereas low-carb beers generally have between 2.5 and 5 grams. Beware of light beers, which can contain nearly 10 grams of carbohydrates—the "light" refers to calories, not carbs—so check the label before you raise a glass.

Wine lovers get a break, carbohydrate-wise, though the numbers can vary depending on the grape varietal. A 5-ounce glass of chardonnay, for example, contains roughly 3.5 grams of carbohydrates, while pinot grigio typically has less than 3 grams. Red wines also vary, although they generally have more carbohydrates per glass than white. Five ounces of Chianti contain about 4 grams of carbohydrates, while a glass of burgundy has 5.5 grams.

While moderate alcohol consumption is linked to some health benefits—including a rise in HDL (the so-called good cholesterol), greater insulin sensitivity, and a reduced risk of developing type 2 diabetes—it can also diminish judgment and increase your chances of ordering another round. That's why I recommend reducing alcohol consumption during Step II of the BioDiet and eliminating it entirely in Step III. If you wish, you can return to drinking alcohol in low to moderate amounts in Steps IV and V.

Common foods with a high glycemic index include:

- white and whole wheat breads, bagels, and muffins;
- white and brown rice, rice crackers, and rice milk;
- white and whole wheat pasta;
- rice- and wheat-flour noodles (e.g., udon);
- processed breakfast cereals, instant oat porridge and rice porridge/congee; and
- sweet potatoes, pumpkin, and potatoes, no matter how they're prepared.

As a general rule, the more cooked or processed a food is, the higher its glycemic index. And the riper a fruit or vegetable, the higher its GI. The more fiber or fat in a food, the lower the number.

I'm often asked whether there aren't health benefits to eating starchy vegetables like potatoes, corn, and peas. While it's true these foods contain important nutrients, including vitamins and minerals, those benefits and many more can be found in low-starch vegetables like kale, brussels sprouts, Swiss chard, rapini, bok choy, and others. These can be steamed or pan-fried in butter, olive oil, or coconut oil. Add some soy sauce, a little sesame oil, and pepper, and you might be surprised at how great they taste. Grated cauliflower that's been sautéed in butter is a great alternative to rice, and boiled cauliflower that's been well drained and mashed with cream, butter, and Parmesan is a nutritious replacement for potatoes.

Notice that I'm suggesting you use a lot of fats in your preparation of these low-starch vegetables. That's because it is essential that you replace the calories you lose by eliminating sugar and starch from your diet with calories from healthy fats. This change significantly increases your metabolism, allowing you to lose weight and inches more quickly.

Decreased sleep
or poor sleep quality
increases hunger and
appetite and decreases
glucose tolerance and
insulin sensitivity—hence
the growing number
of studies that link sleep
loss with obesity.

So, during this second week of Bio-Preparation, eat the fat on meat and the skin on chicken and fish, use lots of butter on vegetables and olive oil on salads, and put whipping cream in your coffee and, yes, even in your tea. And I encourage you to give even those foods you might not be fond of (hello, brussels sprouts) another try. Variety is not only the spice of life, it is also the secret to staying healthy—so long, of course, as those foods are BioDiet-friendly.

Taking Stock: Week 2 Wrap-up

The end of week 2 is another opportunity to evaluate how you're feeling both physically and mentally. Do you have less energy than usual or are you raring to go? Are you feeling more in control of your health or are you overwhelmed by all the changes? Are you starting to get a handle on what to eat and when, or are you having food-security issues? Some people I've counseled tell me that the end of Bio-Preparation is when they begin to experience food boredom. If you do, find a good keto cookbook if you don't already have one. There are lots available, and you'll find recommendations at BioDiet.org as well as a list of websites that offer recipes and other useful materials and information.

Don't forget to journal your experiences, along with a new round of measurements. And, once again, don't be disappointed if you haven't lost weight or inches. As I mentioned earlier, significant changes aren't expected until Step III, which is where we're headed now.

The Glycemic Index

The glycemic index, or GI, of a food is determined by its effect on raising a person's blood-sugar level two hours after consumption. Ranked on a scale of 0 to 100, foods with a value of 70 or above are considered to have a high GI, whereas those with a rating of 55 or less are considered to be low GI. Medium-GI foods are everything in between. Since rising blood glucose increases insulin secretion to allow cells to absorb blood sugar, GI is also a proxy for blood-insulin levels. Foods with a high GI cause high levels of insulin secretion, which, as we know, can lead to insulin resistance, a mainstay in the Axis of Illness.

Recently, a more practical reference, glycemic load (GL), has also come into use. GL factors in both the glycemic index of a food and the amount of it we consume, making GL a better predictor of the actual increase in blood glucose that results from eating the average serving size of a particular food. The formula for determining GL is simple:

GL (average serving size) = grams of "available carbohydrate" × GI
where: "available carbohydrate" = total carbohydrates − fiber.

So, for example, watermelon has a glycemic index of 75, and the average serving—approximately one cup diced or balled—contains 11 grams of available carbohydrates. That's 75 × 11 ÷ 100 = 8.25. Foods with a GL of 10 or less are considered to be low GL, while those with a rating above 20 are high GL. Medium-GL foods comprise everything in between.

Since most discussions, however, still refer to GI, I use that term throughout the book. For a more comprehensive list of foods and their GI and GL, visit BioDiet.org or check appendix D.

7

Step III: Bio-Adaptation
Wheels Up

Intelligence is the ability to adapt to change.
**STEPHEN HAWKING, PHYSICIST AND
COSMOLOGIST, 1942–2018**

YOU'VE GRADUATED TO Step III. Nice work. I know it may have been somewhat of a challenge, but over the last week or two, you've accomplished a lot. You've curbed your sugar and starch consumption, cut back on alcohol, introduced some healthy supplements, and prepared your body for the metabolic conversion to nutritional ketosis that's at the heart of the BioDiet. My hope is that you've also started to change your relationship with food. It was around this point in my own BioDiet journey that I realized I was no longer thinking of food as just something I ate to assuage my hunger but as the most important contributor to good health. According to a recent report on the global burden of disease published in the journal *The Lancet*, poor diet generates more disease than physical inactivity, alcohol, and smoking combined!

Poor diet generates more disease than physical inactivity, alcohol, and smoking combined!

Over the next 14 days or so, you will put into practice everything you learned in part 1 of this book by focusing even more on those elements that allow you to shed weight, increase your energy levels, and, most importantly, help reduce your risk of developing chronic disease. I can't promise Step III will be a breeze, but I can assure you that it will help lay the foundation for a leaner, healthier, happier you.

Bio-Adaptation: An Overview

Some think that ketogenesis—the formation of ketone bodies—is somehow abnormal for humans, but the truth is that it's a natural state that we've all experienced. Newborn babies are in a state of nutritional ketosis and remain that way while they breastfeed. Many researchers have concluded that it is the ketones in breast milk that are responsible for some of the positive effects associated with breastfeeding, including improved learning and development outcomes and better protection from disease, although this has yet to be scientifically confirmed.

Over the course of the next two or so weeks, you will journey back to that state of nutritional ketosis by shifting your metabolism from one that burns sugar as its primary fuel to one that burns fats. It is a more significant transformation than you might think, and it can be more difficult for some than others. That's where this chapter comes in. I believe that by having a better understanding of what happens to our bodies during Bio-Adaptation, we can mitigate and even avoid some of the side effects associated with this metabolic shift. In other words, I'm here for you.

Let's get things started with answers to the questions I'm most often asked.

How long does Step III last?

Although the time required to adapt to fat burning varies from person to person, I recommend sticking with Step III for at least two weeks. By then, you should be adequately keto-adapted and can move on to Step IV, Bio-Rejuvenation.

What changes will I be making in my diet?

During Step III, you will reduce your carbohydrate consumption to 20 grams or less per day and increase your fat intake to compensate. You will also introduce foods and supplements that help with any side effects of keto-adaptation.

What do I need to buy?

- Several Tetra Paks or cans of vegetable, chicken, or beef bouillon (not reduced salt)
- Olives and/or pickles (with little or no sugar)
- Pork rinds (chicharróns), fried in tallow if possible
- Magnesium supplements and/or Epsom salts
- Ketostix (available over-the-counter from your pharmacist)

What else do I need to do?

In Step II, I recommended you keep a record of any changes you experience on the BioDiet. This is particularly important in Step III, as your body adjusts to nutritional ketosis. By recognizing side effects and addressing them early, you can minimize any discomfort. The information might also help your physician determine if there are changes required to any medications they have prescribed for you.

You'll want to continue to use your calendar to help manage your schedule and ensure you reserve time throughout the week to shop for fresh food. Feel free to exercise, though you may find you have less energy during this Bio-Adaptation step. (I discuss exercise in more detail in chapters 8 and 9.) Walks after meals are still important, and they give you the chance to think about how you'll reward yourself when you've completed Step III.

What physical and psychological changes can I expect during this step of the BioDiet?

You will likely experience significant weight loss in Step III. Although some of this can be attributed to the short-term diuretic effect (water loss) of the BioDiet, you should also start to notice a reduction of fat around your midsection as your body begins to tap into its fat stores. This is a more meaningful measure than weight loss because it focuses on those areas where fat deposits pose the greatest health risk.

Over the course of Bio-Adaptation, you should begin to experience more consistent energy levels, improved mental focus, and reduced inflammatory pain. However, you might also experience symptoms of what's commonly called the "keto flu," particularly if you've skipped Bio-Preparation or your typical carbohydrate consumption has been high.

What is the "keto flu" and how do I help avoid it?

Reactions to the metabolic conversion that occurs when you're adapting to a ketogenic diet can be quite varied. Some people sail through it with little more than a mild headache, while others experience influenza-like symptoms that can include achiness, dizziness, irritability, brain fog, diarrhea, stomach pains, and nausea. Some people also report a metallic taste in their mouth or bad breath, particularly in the morning.

Although unpleasant, these side effects are the result of a positive metabolic change and should disappear as your body adapts to burning ketones. The MCT oil I recommended in Step II has helped pre-adapt you to the elevated ketone levels. I advise you to continue taking it, as well as ensuring you stay well hydrated. I also

Side effects are the result of a positive metabolic change and should disappear as your body adapts to burning ketones.

strongly recommend a cup of salted broth once or twice a day. I talk later in this chapter about the role salts play in combating many of the symptoms of keto-adaptation and, more generally, in keeping our bodies functioning optimally.

Bio-Adaptation Basics: Changing Fuel

Until now, the carbohydrates we consumed in excess have produced more than enough glucose for the brain and other tissues—hence our high blood sugar, high insulin levels, and the weight gain and

The Harper High

When my wife and I adopted the BioDiet years ago, we were both keen to lose some weight and enjoy the health benefits of a ketogenic diet. The first few carbohydrate-free days were relatively easy. But by day 3, we were experiencing headaches, difficulty focusing, and low energy. The side effects over the next couple of days were augmented by those flu-like symptoms and also by poor sleep and some serious sugar cravings. I began to wonder whether the destination was worth the journey. That all changed for me on day 5 when I snapped awake feeling alert, optimistic, and full of energy. The people I've counseled since jokingly call this the "Harper High," and though not everyone experiences it as quickly or suddenly as I did, almost everyone reports improved mood, a feeling of greater well-being, and better mental clarity. My favorite description, though, is from the woman who told me it felt as if someone had pulled a decade's worth of cotton wool from her brain. In terms of incentives, I'd say it doesn't get much better than that.

disease that results from both. I've explained how we can help address these problems by reducing our carbohydrate intake, a process we began in Bio-Preparation. If, however, we want to trigger the production of ketones in our bodies (and trust me, we really do), it isn't enough just to cut back on carbohydrates: we need to all but eliminate them from our diet, at least for now.

As you recall, there are three types of macronutrients: proteins, fats, and carbohydrates. These provide us with the calories we need and are the organic building blocks for normal tissue growth and repair. Of the three, only two—proteins and fats—contain components that are essential, meaning that we need to consume them because we can't manufacture them ourselves. Carbohydrate is *not* an essential nutrient, so we can survive and, in fact, thrive without it. That said, some carbohydrates, like fiber, are beneficial in that they help control blood sugar, lower cholesterol, and regulate bowel movements.

As for the two essential macronutrients—proteins and fats—maintaining the right balance between them is important (roughly three calories of fat for every calorie of protein). Though no healthy upper limit of protein consumption has yet been identified, the National Academy of Medicine in the United States recommends a minimum of 60 grams of protein per day for the average adult female and 71 grams for the average adult male. However, most of us are eating twice that amount. While this level of protein consumption might be fine for some, for others it could lead to kidney problems, digestive issues, or weight gain. Most research points to a reasonable upper limit of about 25 percent of calories from protein. Much more than that, however, will result in excess amino acids being converted to blood glucose, which will prevent you from maintaining a state of ketosis. So, if our tolerance for protein is limited and we've all but eliminated carbohydrates from our diet, where will our calories come from? Nutritious, tasty fat.

Carbohydrate is *not*
an essential nutrient, so
we can survive and,
in fact, thrive without it.

Fat's Bad Rap

If you grew up in a household where eating fat was frowned upon, you're not alone. During the 1980s and '90s, the low-fat craze changed not only the way we ate but also what we *believed* about what we ate. Fat was considered unhealthy and unpalatable (never mind that it lends much of the flavor to foods). Skim milk was the new normal, butter was replaced by margarine, and the dreadful fat-free frozen yogurt was born. That decades-long demonization of fat continues to this day, which helps explain why so many of us have trouble getting our heads around the fact that fats are good for us. In fact, they are the most energy dense and, in the case of saturated fats, cleanest burning fuel we can put in our bodies. That's why the BioDiet consists of about 70 percent fat, 25 percent protein, and 5 percent carbohydrate.

What does that look like on a plate? For the 5 percent carbohydrate, think fiber-rich foods including celery, artichokes, broccoli, brussels sprouts, berries, almonds, flax, and chia seeds (so much for the misguided notion that ketogenic diets don't contain enough fiber). The 25 percent protein can come from fish and shellfish, poultry, meats, and unsweetened soy products. As for the 70 percent fat, I recommend avocados and macadamia, olive, avocado, and coconut oils; butter, cheese, cream, eggs, nuts, seeds, and olives (see appendix A for a complete list). These healthy fats are as essential to our survival as water and salt.

The Stuff of Life, Part I: Water

All vertebrates, including humans, are descendants of organisms that once lived in the ocean. Sure, we humans eventually learned to walk around on our two feet, but we hung on to seawater as our basic bodily fluid, which is why we don't survive very long without water. Why, you ask, am I talking about evolution in a book about nutritional ketosis? Because getting enough water is crucial to the BioDiet. In response to decreased carbohydrate consumption, the body draws on glycogen stores for energy. However, one of the inefficiencies of glycogen storage is that it requires water—a lot of water. For each gram of glycogen that's stored, 3 to 4 grams of water are along for the ride. So, as we deplete our glycogen stores during Bio-Adaptation, we lose the water that was stored with it. That can amount to up 2 kilograms of water, or roughly 8.5 cups. Feeling parched?

Not only that, reduced carbohydrate consumption also results in reduced blood sugar and insulin. Since insulin promotes water retention by inhibiting sodium excretion, as insulin levels drop, the kidneys excrete sodium and water follows it, which is why some of the initial weight loss on the BioDiet can be attributed to that diuretic effect. Making sure you stay hydrated not only helps the kidneys do their job, it also helps prevent the headaches, confusion, muscle cramps, and general irritability that can result from dehydration. It's important to note that the diuretic effect of a ketogenic diet tends to reduces blood pressure, at least in the short-term. This is great if you have high blood pressure; however, if you are already taking medication to reduce it, you should be carefully monitored by your physician to ensure it doesn't drop too low.

The Stuff of Life, Part II: Salt

That seawater we just discussed? It's full of salt, so in addition to ensuring you get adequate hydration, it's important to get the salts right. Among other key functions, salts form electrolytes—substances that ionize to become particles with positive or negative charges—when they're dissolved in water. This is vital for nerve and muscle function. When we think of salt, we usually think first of table salt—sodium chloride—so I'll start the discussion there.

Sodium

Like fat, sodium is much maligned in our present, conventional food recommendations. The reasoning behind this is that excess sodium increases water retention, which increases blood volume, which, in turn, raises blood pressure. However, sodium is an essential nutrient, and recent studies have shown that the present recommended amounts—between 1,500 and 2,300 milligrams per day for younger adults, and even lower for those over the age of 50 or under the age of 8—are insufficient and might be negatively affecting our health. By addressing the root cause of salt retention—high insulin levels—the BioDiet promotes the body's natural method for dealing with high sodium levels, which is to become thirsty, drink fluids, and allow the kidneys to secrete the excess sodium.

Sodium chloride is used in large quantities in processed foods like bread, which you might be surprised to know is the leading source of salt consumption in the North American diet. Fast food also contains enormous amounts of salt, which is why it is often accompanied by a giant-sized beverage, usually a sugar-sweetened one. Since the BioDiet is about as far from highly processed fast

It's important to get the salts right.

food as you can get and, of course, features no high-carbohydrate bread, it's important to ensure you're getting enough salt. Like dehydration, symptoms of insufficient salt intake can include headaches, dizziness, nausea, muscle cramps, and even heart palpitations. Furthermore, studies have shown that low sodium intake not only increases insulin levels, but it also increases the synthesis of fatty acids in the liver, which can contribute to the development of fatty liver disease. So, during Bio-Adaptation, be liberal with the salt-shaker and augment your consumption with bouillon and other salty foods such as olives and pickles.

It's also important to note that salt is where we get most of our dietary iodine, an essential element that helps the thyroid gland regulate metabolism. While gourmet salts like pink Himalayan, French sel gris, and fleur de sel are tasty (and pricey!), they don't often contain iodine. So be sure to read the label and choose one that does, or opt for your run-of-the-mill iodized sea salt. Your body and your wallet will thank you for it.

Sodium chloride isn't the only important salt for our bodies. Other dietary minerals, like potassium, magnesium, and calcium, are also essential for good health. Unfortunately they are three essential nutrients we don't get enough of.

Potassium

Inside your body, potassium functions as an electrolyte that helps regulate fluid balance, nerve signals, and muscle contractions—including those of the heart—and it may reduce blood pressure, protect against strokes, and help prevent kidney stones and osteoporosis. Fortunately, the BioDiet is rich with foods high in potassium, such as avocado, leafy greens (including spinach, kale, and beet greens), cruciferous vegetables (such as broccoli and brussels sprouts), along with fish, meat, and nuts, especially almonds.

According to Connie Weaver of Purdue University, the average American only gets about half of the potassium they need, and only 3 percent of people meet the minimum daily requirements. Symptoms of low potassium are generally muscle related, like twitching and cramping, but can also include heart palpitations. If you experience any of these, have a sugar-free sports drink that contains potassium or, better still, eat an avocado. You can also take potassium supplements.

Magnesium

Magnesium regulates more than 300 of the body's cellular reactions, and most of us aren't getting enough. If you're among that group, you might have muscle cramping while adapting to the BioDiet, especially at night or after exercise. You might also experience nausea, fatigue, constipation, and/or weakness. If your magnesium deficiency is chronic, it can aggravate cardiovascular disease, diabetes, osteoporosis, and migraine headaches. Magnesium-rich foods include spinach, Swiss chard, avocados, chocolate, almonds, soybeans, and seeds (pumpkin, sunflower, sesame, hemp, and chia). A pleasant way to increase your magnesium levels is to add a couple of cups of Epsom salts to your bathwater. Studies have shown that a single salt bath can increase blood magnesium levels by about 10 percent, which increases to 40 percent if repeated daily for a week.

Since magnesium also has anti-inflammatory properties, it's a great post-workout option.

You can also take magnesium in pill form. Roughly 300 to 400 milligrams per day is fine. However, if you have kidney problems or are taking any medication, consult your physician first. And keep in mind that some forms of the mineral, like magnesium hydroxide (milk of magnesia), have a significant laxative effect, while others can result in nausea and abdominal cramping. You can reduce these side effects by taking magnesium with meals or ensuring your supplement also contains calcium, which slows the absorption of the magnesium and reduces the laxative effect. If you are experiencing abnormal or persistent constipation, magnesium hydroxide can do the trick.

Calcium

This is the most abundant mineral in the body. When combined with phosphate, it makes up most of our hard tissues, like teeth and bones. It is also important for our nervous system, heart function, and muscle contraction. Our bodies regulate the amount of calcium in our blood so that, like the other electrolytes, it stays within a narrow range. When calcium levels are low, our parathyroid gland produces a hormone that tells our bones to release more calcium into the blood. Over time, this can compromise bone health, which is why ensuring you get enough calcium is crucial. The BioDiet consists of foods rich in calcium, including dairy products, nuts, soy, broccoli, and spinach. Adults need about 1,300 milligrams of calcium per day, so if you don't think you're getting enough—and recent studies indicate that only half of us are—consider taking a calcium supplement. Find one that also contains vitamin D, as that helps with absorption, but be sure not to overdo it: excessive calcium intake can result in kidney stones. One way to avoid that is to add lemon or lime juice to water or foods. The citric acid in it will bind excess calcium, which is then removed in the urine.

What to Eat: Sticking with the Plan

The better prepared you are for your meals, the more likely you are to stick with the BioDiet. If you open up the refrigerator to find no nutritious food, you are a lot more likely to return to your old ways of eating. During Bio-Adaptation, it takes very little carbohydrate to halt your body's transition to burning ketones, so unless you're prepared to start the metabolic conversion process all over again, it's important to keep your carbohydrate intake to a minimum. Planning and preparation are the keys to that. Here are some meal suggestions to help you stay on track.

Exercise during Bio-Adaptation

If you're one of those people who prides themselves on working out hard on a regular basis, you might not like what I'm about to say. Exercising during Step III isn't necessarily advisable. That's because, during Bio-Adaptation, your body already has a significant job to do—shifting metabolism from burning one type of fuel to burning another. Strenuous exercise can complicate the process and make it difficult to determine whether any achiness, dizziness, headaches, or intestinal problems you might be experiencing are symptoms of keto flu or the result of an overly ambitious workout. So listen to your body, because there's a good chance it's telling you to go easy. Continue to walk, and perhaps include some yoga or stretching in your day to help address any stiffness or muscle soreness. As for pumping iron or 10K runs, you'll have plenty of opportunity to get back to that in Steps IV and V. Besides, it's beneficial to take a break from your workout routine once in a while to let your body recover and rebuild. Two weeks is a good length of time for that.

Breakfast

In Step II, I offered some morning meal ideas, and you probably came up with some low-carbohydrate options of your own. You can continue to eat those as long as you're not consuming any grains (even whole grains), cereals, bread, bagels, muffins, bakery products, or anything made with flour, corn, or other starch. Fruit is also off-limits for now. (We re-introduce berries in Step IV.)

Before you eat anything, however, start your morning by consuming two full glasses of water. Then, if possible, wait 20 minutes before you eat or have your tea or coffee. You may remember from Step II that eating in the morning isn't essential to the BioDiet. That's still true. However, there are a few things you may want to consider about a morning meal during Step III. The first is that because your blood glucose is very stable on the BioDiet, you might not experience the sensation of hunger associated with low blood sugar. This means it can sometimes be hard to tell when it's time to eat. Some people on the BioDiet report that their need for food is characterized by decreasing energy levels or even sleepiness rather than hunger. I experience it as a kind of winding down, which lets me know that I need to eat soon or risk being found asleep on my keyboard. The second consideration on the breakfast front is that it can be tough, if not impossible, to find very low-carbohydrate food options at your local coffee shop or convenience store. So, if you typically leave the house for work before eating, I suggest you pack a hardboiled egg, some beef jerky, a can of tuna, or a small handful of pecans or Brazil nuts, which are low in carbohydrates. That should tide you over till lunch.

Mid-Morning

As we've learned, Step III of the BioDiet is not just about eliminating carbohydrates; it's also about greatly increasing fat consumption. This can take the form of avocado, tasty chicharróns (fried pork rinds), or high-fat dairy. One of the best and most pleasurable ways for me to get fat into my diet is with a caffe latte made with cream. It's called a breve latte, and once you taste it, you'll never go back to one made with low-fat milk.

I know that drinking cream can seem counterintuitive if you're trying to lose weight, but remember that you are no longer eating carbohydrates, which means you need to get your calories elsewhere. By transitioning to fat as your energy source, you're not only

The Truth about Fructose

It always surprises me when someone says they don't want to eat sugar substitutes because they're "full of chemicals" and bad for our health. In fact, it is sugar—also a chemical—that I, and many others, believe is the real problem.

As you'll recall, sucrose (table sugar) is half glucose, half fructose. Fructose, which is the main ingredient in high-fructose corn syrup, is often added to processed foods—including energy bars, breakfast cereals, and sugary drinks—because it is cheap and enhances taste. But unlike glucose, which is metabolized in cells throughout the body, fructose metabolism occurs almost entirely in the liver. The by-products of this process are inflammatory chemicals and fat molecules, which negatively affect many of our organs and can lead to serious chronic conditions.

Liver
- may cause non-alcoholic fatty liver disease
- increases the production of uric acid, which over time can lead to gout

burning the fat you consume, you're also using up the fat stored in that spare tire around your waist.

Lunch

If you're at home during the day or you have a microwave at work, consider eating leftovers from the previous night's dinner. They're often more flavorful than what you can find at your local lunch spot, not to mention a good deal less expensive. If you want something

Kidneys

- results in renal vasoconstriction, which reduces blood flow to the kidneys
- has been linked to hypertension, which results in excessive pressure on the kidney's microscopic filtering structures

Brain

- may play a role in opioid dependence by triggering a reward response that leads to continued cravings, similar to those experienced by drug addicts

Blood vessels

- causes damage to the cells that line the interior surface of blood vessels, leading to arteriosclerosis

Studies in animals indicate that, in addition to the above, dietary fructose increases blood triglycerides, promotes insulin resistance, and significantly increases the production of visceral adipose—the fat that lies deep within the abdominal cavity—increasing our risk for cardiovascular disease and type 2 diabetes, breast cancer, and gallbladder issues.

By transitioning
to fat as your energy
source, you're not only
burning the fat you
consume, you're also
using up the fat stored
in that spare tire
around your waist.

fresh, a salad may be your best choice, accompanied by some sort of non-breaded protein, like chicken, shrimp, or tofu. Soups are also a good option, especially cream-based ones, so long as they aren't thickened with flour, or contain rice, pasta, legumes, potatoes, or other starchy vegetables.

Mid-Afternoon

From a consumer's standpoint, one of the best things about the growing popularity of ketogenic diets is the proliferation of ketogenic-friendly protein bars. These are especially useful on days when you're on the go or just need a little extra energy. Choose one that contains the least net carbohydrate (see appendix C: Nutrition Labels) and the most palatable sugar substitute, which for me tends to be erythritol. (For a complete list of sugar substitutes and their various pros and cons, see appendix B.) However, these bars do contain carbohydrates and, because they're sweet, they can aggravate sugar cravings, so it's best not to indulge in them too often. Save them for emergencies when you don't have access to other options and feel the need to eat.

Dinner

Dinner is your chance to get a variety of nutrient-dense foods into your diet—think watercress, Chinese cabbage, chard, beet and collard greens, spinach, endive, and kale. Nutrient-dense low-carbohydrate proteins include pork, beef, lamb, chicken, salmon, tuna, and shellfish. Tofu, eggs, and high-fat cheese are good alternatives for vegetarians. Anything from the grocery list in appendix A is fair game.

If you're still hungry at the end of your meal, have a small piece of high-fat cheese, but don't overdo it. Going to bed full could disrupt your sleep, and if the foods contains carbohydrates, they will raise your blood-glucose levels, interfering with the release of growth hormone, which is responsible for the repair and renewal of tissues and promotion of lean body mass. If you have a sweet craving that won't subside, have a sugar-free candy.

As I mentioned earlier, you can help address most symptoms of keto flu with a cup or two of salted broth. Hot water with lemon or an herbal tea before going to bed also increases hydration levels.

Checking Your Progress: Number Crunching

Once you have completed Bio-Adaptation (and keep in mind that, for most people, this will be a two-week step, but it might take longer), your metabolism has been reprogrammed to burn fat. This is a good time to take stock of your progress. You can do this in a few different ways.

Do You Need to Count Calories and Carbohydrates?

In a word, no. This is yet another advantage of the BioDiet.

Counting *calories* often leads to calorie restricting, which is something I don't want you to do. Stick with the foods on the grocery list and, as physician and ketogenic clinician Dr. Eric Westman of Duke University advises, "eat when you are hungry, stop when you are full."

The same goes for counting *carbohydrates*. Eating only those foods on the list should keep you in nutritional ketosis, but if you're worried you're not in ketosis, simply cut back on your carbohydrate consumption. If you're determined to count carbohydrates, you certainly can, but to do it with any accuracy, you need to know the nutritional values for all of the foods you eat and the exact weights and/or volumes. I recommend the Keto Diet Calculator provided by the Charlie Foundation for Ketogenic Therapies (see KetoDietCalculator.org).

Counting Ketones

The goal for Step III is to transition to a state of nutritional ketosis. Ideally, that is a blood-ketone level above at least 0.5 millimoles per liter (mmol/L) to about 3.0. So how do you determine if you've made it? Good question. The body produces three different ketones, and their production and metabolism vary considerably between individuals and also over time. Therefore, your ketone "status" does not necessarily accurately reflect your production and use of ketones. The best way to ensure you achieve nutritional ketosis is to limit your carbohydrate intake as much as possible—as you've been doing in this step. But if you do want to get some measure of your ketone status, there are three ways to test it.

- **Urine:** The Ketostix recommended in the shopping list are the least expensive, easiest, and most common way to test your ketone levels, and the test might indicate if you are in nutritional ketosis. Urinate on the strips first thing in the morning, before you eat or drink, and look for changed color on the strip. It's important to note, however, that the results aren't always accurate— you can be in nutritional ketosis even when the Ketostix shows otherwise and vice versa. The Ketostix test is just a ballpark estimate, which is why I suggest you stick with the foods I've recommended and measure your success based on fat loss, improved mental clarity, increased energy, and decreased hunger.

- **Breath:** A second method of determining your ketone levels is to use a ketone breath monitor, which measures the amount of acetone in your breath. However, the accuracy of the results will depend on the quality of the breath monitor, which can cost upwards of $250.

- **Blood:** By far the best way to measure your ketone levels is by sampling your blood in much the same way diabetics do to measure their blood-glucose levels. A lancet is used to pierce the skin and a drop of blood placed on a test strip, which is input into a device that measures beta-hydroxybutyrate ketone bodies. While blood-ketone monitors are more accurate and tend to be less expensive than breath monitors, the cost of the test strips (a dollar or more each) can add up over time.

And speaking of measurements, I saved the best ones for last!

Measure Up

The end of Bio-Adaptation is a great time to retake the measurements you completed back in Steps I and II (weight, waist/hips, BMI, and RFM). Jot down your new numbers and compare them to your previous measurements. There's a good chance you've lost between 5 and 15 pounds, but more importantly, you've likely lost an inch or two off your waist and are feeling better than you have in years. At this point, many people also report decreased musculoskeletal pain, a significant increase in energy during the day, and a better night's sleep. If you snore, your partner might happily notice that it is abating. Your gut bacteria are adjusting too and producing a lot less gas. Once again, take note of any improvements you're experiencing as well as any concerns you might have. This information helps keep you on track as you move into the next stage of your BioDiet journey.

Reward Time!

Now that much of the heavy lifting is over and you are well on your way to better health, it's time to deliver on whatever fantastic reward you promised yourself. Celebrating your success not only acknowledges all the hard work you've done, it also paves the way to the important next step, Bio-Rejuvenation.

8

Step IV: Bio-Rejuvenation
Winding Back the Biological Clock

The greater the obstacle,
the more glory in overcoming it.
MOLIÈRE, FRENCH PLAYWRIGHT,
ACTOR, AND POET, 1622–1673

WELCOME TO STEP IV of the BioDiet. If you're like me, you found these past weeks both rewarding and challenging. You've been asked to change not only what you eat but also how you think about food. You've eliminated sugar from your diet along with the foods your body converts to sugar. And you've replaced most carbohydrates with healthy, satisfying fats; a variety of proteins; and delicious, nutrient-dense vegetables. The heavy lifting is over—I hope you gave yourself a pat on the back and made good on that reward you promised yourself for the completion of Step III.

Speaking of rewards: the number on your bathroom scale should be on its way down, and your energy level on its way up. Stay the course with Step IV and you'll continue to lose weight, increase your vitality, and greatly improve your overall health and happiness.

Bio-Rejuvenation: An Overview

I call this eight-week step Bio-Rejuvenation because this is time when those I've counseled begin to look and feel so much better that it seems as if they're getting younger by the day. In addition to continued weight loss, people report improved sleep and more alertness when they awaken; many tell me they experience a more stable and improved mood, increased energy, and a more positive feeling about life in general. Sounds good, right? Read on.

How long does Step IV last?

Eight weeks, during which time your body continues to progressively adjust to its new fat-burning metabolism.

What changes will I be making in my diet?

During Step IV, you can reintroduce limited amounts of some carbohydrates, including a little fruit, and you can drink moderate amounts of alcohol if you want.

What do I need to buy?

There are some additional foods, including beneficial spices, that you can purchase to help make your meals more varied and satisfying during this step (see appendix A).

What else do I need to do?

Much of the challenge in Step IV is psychological: staying with the program. Doing so will ensure that you continue to shed fat and improve your health. You'll notice the former, but the latter is what's really important. With that in mind, I recommend that if you haven't yet adopted an exercise program, you begin one in Step IV. If you are already exercising, consider ramping up your routine. Later in the chapter, I explain why the Bio-Rejuvenation step is an excellent time to do that.

You can continue to test your ketone levels throughout the next eight weeks, however, if you're using Ketostix, don't be surprised if they show reduced levels of ketones during Bio-Rejuvenation. As your body adapts to the BioDiet, it burns ketones as quickly as it produces them, which results in fewer ketones being excreted in the urine. If you are using a breath or blood monitor to determine your level of ketosis, a reminder that the target range should be between 0.5 and 3.0 mmol/L. That said, please don't obsess about the short-term numbers. It's the long-term *results* that are important.

What physical and psychological changes can I expect?

During Step IV, you should lose 1 to 2 pounds per week, although this varies from person to person and does not always manifest as a steady decline. More important than weight loss, however, is the decrease in your waist measurement, since, as we know, this correlates more strongly with overall health.

As your blood glucose stabilizes, your sleep should improve, which further promotes weight loss, along with the repair and renewal of tissues. And by continuing to keep inflammation in check, you can expect less joint and muscle pain. Many people report improved skin quality, due to reduced inflammation and increased dietary fat, so don't be surprised if your friends and colleagues tell you how great you look.

Bio-Rejuvenation: The Basics

Over the next two months, your blood sugar will further stabilize and your insulin sensitivity will improve. Your clinical markers for chronic disease will also improve, especially lower triglycerides and higher HDL (good cholesterol). Now you can slowly increase your carbohydrate consumption from less than 20 grams a day to as much as 50 grams. Foods you can consider working back into your diet include:

- small amounts of berries, especially strawberries, raspberries, blackberries, and blueberries;
- the occasional small slice of apple, pear, or kiwi;
- more nuts and seeds (no more than a handful a day, and avoiding those higher in carbohydrates, like cashews);
- more cheese, especially those high in fat (cheddar, gouda, Swiss, Parmesan, Brie or Camembert, blue, Jarlsberg, goat cheese, feta, and mozzarella);
- spaghetti squash, summer squash, snow peas, and eggplant;
- nut flours; and
- wine, spirits, or very low-carbohydrate beer, in moderate amounts.

A few words of caution: be careful not to increase carbohydrates too quickly or to exceed the 50-gram daily maximum or you can knock yourself out of ketosis. If you are a type 2 diabetic, you are probably best staying with carbohydrate restriction closer to that 20-gram mark until your physician advises otherwise.

A good approach is to eat small amounts of any new foods and not add them all at once. Start with one or two changes per week and monitor your morning body weight and waist size. If you find your weight loss stalling or, worse, starting to rise, cut back your carbohydrate consumption to Step III levels for a week.

What to Eat: The Benefits of Balance

The reintroduction of certain foods in Step IV can present some challenges. To help you manage this important but at times trying stage, here is a breakdown of the best times during the day to incorporate these additional carbohydrates into your diet.

Breakfast

Morning is ideal for introducing berries for several reasons. First, they taste great when combined with high-fat yogurt and a sprinkling of toasted seeds. More important, however, morning is when your insulin sensitivity is at its highest, which makes your body particularly effective in converting the carbohydrates in the berries to energy rather than storing it as fat. Higher insulin sensitivity also reduces the amount of insulin you need to return your blood-glucose levels to normal after eating, which helps moderate blood-insulin levels. Begin by limiting your berry consumption to 2 ounces (a small handful) per day, no more than a few times a week.

Mid-Morning

Over the last few weeks, you've been consuming a maximum of 2 ounces of nuts per day. Now that your body has successfully transitioned to burning fat, you can double this, but don't forget to account for nut butters in your tally.

As always, choose nuts that are highest in fat and lowest in carbohydrates, like Brazil nuts, pecans, macadamia nuts, and walnuts. Pumpkin seeds are also a great low-carb, high-protein snack, but try to resist eating them by the handful—the carbohydrate count adds up quickly. Peanuts, too, are relatively low in carbohydrates, but these legumes contain proteins that can cause stomach issues and bloating in some people. Although almonds have more carbohydrates than some other nuts, two-thirds of that is healthy dietary fiber. Almond meal or flour is great for use in baking, but keep track of how much you are using since it, too, goes toward your maximum daily carbohydrate intake.

Nut lovers are almost always disappointed to learn that cashews are high in carbohydrates and low in the dietary fiber that would offset the carbs, so go easy on them. And avoid any nuts or seeds cooked in processed vegetable oils, which can contain significant amounts of the pro-inflammatory omega-6 fatty acid. If you're going to eat pistachios, buy them still in their shells; you'll eat them slower and consume fewer.

Lunch

Preparing lunch at home is the easiest way to stay on course with the BioDiet, so it's worth spending a bit of time on your weekend making food to take with you during the workweek. I like to do this on Sundays. I'm a big football fan (go Titans!), and we have a TV in the kitchen, so I just let the games roll on while I cook. There are many great ketogenic and low-carbohydrate cookbooks on the market, as well as an abundance of websites with tasty recipes, including good lunch options. You'll find a list of the recommended books and sites at BioDiet.org.

Mid-Afternoon

Cheese can be a good source of fat and protein, which makes it a great option if you find yourself running out of steam by midday. Most cheeses are low in carbohydrates, which is why you can double your consumption in Step IV. As always, choose high-fat varieties and avoid low-fat or fat-free products since these can contain significantly more carbohydrates than the full-fat versions. Low-fat and fat-free cream cheeses, for example, contain 8 grams of carbohydrates per 100 grams, which is almost twice as many as the 4.1 grams found in full-fat cream cheese. Fat-free sour cream, at 16 grams of carbohydrates, has nearly six times that of the full-fat version, which contains 2.6 grams.

Dinner

Variety is essential to the BioDiet for a couple of reasons. Obviously, it's important that you get all the vitamins and minerals you need to stay healthy, but there's also a sly reason that I'm encouraging you to eat as many different foods as possible: food boredom— the silent saboteur that can tempt even the most disciplined of us to stop at the donut shop on our way home from work. To avoid food boredom, begin with a meal plan that includes your favorite BioDiet-friendly foods prepared in different ways. This whets your appetite for exploring new ingredients while still ensuring there's always something on your plate that you like to eat. Another strategy is to combine the foods you enjoy with the ones you don't: melted Brie on broccoli; bacon bits and sugar-free maple syrup on brussels sprouts; almond-encrusted whitefish; and toasted sesame seeds and soy sauce on fried tofu. If all else fails, grass-fed butter can make almost anything palatable. Plus, it's packed with vitamin K, which helps with blood clotting, and omega-3 fatty acids, which are anti-inflammatory and help reduce your risk of developing heart disease.

If all else fails, grass-fed butter can make almost anything palatable.

The right seasoning can transform bland foods into tasty ones, and some spices actually improve your health by reducing inflammation. Turmeric is great with eggs, in soups, and with braised or sautéed greens. Combine cayenne pepper, paprika, onion powder, thyme, basil, and oregano to make a Cajun rub for fish, chicken, or pork. Garlic and ginger are also excellent with chicken, shellfish, or in a vegetable stir-fry. And a pinch of cinnamon is a good match for lamb as well as buttered summer squash, which you can introduce in small amounts in this step.

Post-Dinner

Snacking after dinner generally has little to do with hunger. We often do it because we're bored or stressed, or to satisfy a craving. Some eat impulsively to make up for a bad day, while others use food to reward themselves for a good one. Whatever the motivation, after-dinner snacking is often uncontrolled and tends to involve foods with little or no nutritional value. If you do feel the need to eat at night, portion out a small low-carbohydrate snack like cheese or nuts and eat it mindfully. Not only will you consume less, you'll also enjoy it more.

Bedtime

I recommend you refrain from eating after 8 p.m., especially if you're looking to lose weight. Remember that most growth hormone is released in your first hour or so after falling asleep, and the pituitary gland that releases it is sensitive to blood glucose: the higher your blood sugar, the less growth hormone released. Growth hormone promotes lean body composition, so going to bed a little hungry can reduce your blood sugar and maximize the hormone's effects. This not only helps with weight loss but also promotes the repair and renewal of tissues. No wonder it has earned a reputation as the anti-aging, rejuvenating hormone.

Staying the Course

We've talked a lot about *what* to do on the BioDiet but not so much about *how* to stick with it. As anyone who has tried to make a positive lifestyle change knows, the tricky part is to avoid giving in to old habits. This is especially true when it comes to diets. The best estimates suggest that only about 50 percent of us are able to maintain a diet over the long term (more than one year).

What sets apart those who are able to stay the course from those who don't? To find out, I asked my friend and colleague Roger Friesen, who knows a lot about not giving up when the stakes are high. Roger is a sports psychologist who works with Olympic and professional athletes to help them maintain their motivation, despite the sacrifices they must make to achieve their goals and the setbacks they may encounter along the way. The key, he says, is to continue to believe in your ability to succeed, especially when things get tough. Here are some of the strategies he suggests.

- Give yourself credit for recognizing the need to make changes.
- Commit to a regimen that helps you accomplish your objectives.
- Realize that you have the ability to reach your goals.
- Know that not everyone will support your efforts.

In the end, it is your belief in yourself that matters most. Accept your missteps and get back in the game; acknowledge your successes and don't be shy about celebrating them.

My hope is that Roger's advice will not only help you stick with BioDiet, but that it also might prove useful in a broader sense, because losing weight and getting healthy are just two among many ways to improve your well-being.

Maintaining Momentum:
Battling the Dreaded Plateau

Let's say you've greatly limited carbohydrates, have a handle on your wine consumption, and quit snacking at night. Now you're losing weight, feeling healthier, and looking great. Suddenly, though, your weight loss slows or perhaps even stalls. You're super-frustrated. You think you're doing everything right, but the numbers on the scale refuse to budge. Although this is not a common occurrence on the BioDiet, it does happen; weight loss can be as variable as the people experiencing it. Some will lose a steady 1 or 2 pounds per week while others may lose weight in spurts, perhaps slimming down significantly in the first month and then shedding fat more slowly. How do you ensure you're doing everything you can to get the scale moving again? Here are some reasons why your weight loss may have plateaued and what to do about it.

You're Not Eating Enough

Not eating enough is probably the most common reason for slowing weight loss. As we know, calorie restricting slows down your body's metabolism, which hampers weight loss. By eating more calories in the form of fat, you not only boost your metabolism, you also cause your body to burn the fat stored in the spare tire around your waist and hips. The BioDiet encourages you to enjoy full-fat dairy, use olive oil liberally, and savor meat, poultry, and wild-caught fish high in fat. Choose organic animal protein whenever possible, since hormones and pesticides tend to be fat-soluble and collect in higher concentrations in the fat of animals.

You're Eating Too Much

And by that, I mean you are eating too much of those foods that can inhibit ketosis. If your keto testing reveals that you are no longer in a

Calorie restricting slows down your body's metabolism, which hampers weight loss.

state of nutritional ketosis, it's time to reduce consumption of some of the foods you reintroduced in Step IV. These include berries and squash along with additional portions of nuts and cheese.

Eating too much protein can also stall weight loss because the body converts the excess amino acids in the protein to glucose. I recommend you limit your protein consumption to a maximum of 25 percent of your total calories. If you exercise strenuously on a daily basis, you may want to increase your protein intake to between 30 and 35 percent. Keep in mind, however, that a diet containing too much protein may also be linked to kidney problems, though much more study is needed to determine if there is a true relationship.

You're Drinking Too Much Wine

If there's one place that I slip up, it's on wine consumption. That's my kryptonite. Not only does wine contain some carbohydrates, as we learned in part 1, but the alcohol also acts as an appetite stimulant by interfering with the effects of the hormones ghrelin and leptin, which increase and decrease hunger respectively. Since alcohol also impairs judgment, you may find yourself drinking more than the allotted number of glasses per day (one for women, two for men) or giving in to unhealthy food cravings. If you are going to drink wine, it's best to consume it with food, which slows down alcohol absorption and eases the load on your liver. The same applies to spirits or low-carbohydrate beers.

You're Not Drinking Enough Water

Throughout all of the BioDiet steps, I've emphasized the importance of staying hydrated. That's because dehydration can not only negatively impact brain function, blood volume, and heart rate, it can also inhibit weight loss. Water is involved in almost every biological function in the body. When you're not sufficiently hydrated, your metabolism slows down, which can reduce your body's ability to burn calories and to use fat as fuel.

By now, you should have developed the very good habit of drinking water early and often. Seven years in and I still start my morning with at least two glasses of water, and I ensure that I continue to hydrate throughout the day.

You're Cheating

We've all done it—slipped in a cookie or two, given in to the temptation to eat crackers, or raided the kids' cereal cupboard before going to bed. As we know, carbohydrates and sugar, in particular, can be hard to resist, so if you do cheat, don't beat yourself up. Nor should you use it as an excuse to keep eating carbohydrates, especially if you're going to continue eating fats. The combination of the two will not only cause you to regain the weight you've lost, and probably more, it will also compromise your health.

Start by making a note of what you ate along with your reasons for eating it. Did the dessert menu at your favorite restaurant present too much of a temptation? Did you succumb to the potato chips or nachos at the get-together? Or was the smell of fast food on your way home from work irresistible? Knowing your triggers allows you to come up with strategies to prevent you from cheating in the future—perhaps bringing a piece of sugar-free chocolate to the restaurant to enjoy after your meal, eating a low-carbohydrate snack before going to a party, or making sure your refrigerator is well stocked with your favorite BioDiet-friendly dinner foods. You may also find it helpful to write down any physical reactions you experience from consuming carbohydrates. For some, eating carbo-hydrates after not consuming them for a time can result in an upset stomach, gas, or diarrhea. When I've fallen off the wagon in the past, I've experienced aching joints and mild flu-like symptoms as a result of inflammation. It was enough to keep me from doing it again.

Knowing your
triggers allows you
to come up with strategies
to prevent you from
cheating in the future.

You've Reached Your Natural Set Point

Set-point theory proposes that our bodies are programmed to be a certain weight or to store a certain amount of fat, and though that set point can fluctuate, we tend to stay within range of it. Of course, that doesn't square with the world's growing obesity rate or the fact that people tend to gain even more weight after going off calorie-restricted diets. What I have noticed, however, in those I've counseled is that most people lose weight until they reach a size and body composition they're happy with and then naturally adjust their diet to maintain it. Since I lost 27 pounds of fat seven years ago, I've remained at 150 pounds, plus or minus about 3 pounds. I could lose more weight if I wanted to by further restricting my carbohydrate intake, or I could easily gain weight (hello, beer!), but as long as I stay in nutritional ketosis, my weight and body composition remains stable.

You're Gaining Muscle Mass

If you're working out on a regular basis, and that routine involves resistance training (lifting weights, for example, or suspension training), you may be gaining muscle mass at the same time as you're shedding fat. Since muscle tissue is denser than fat, your weight may be holding steady or even increasing, despite the fact that your body is getting leaner. You can check this by calculating your relative fat mass using the formula in chapter 5. Aim for the range considered healthy for your age category, but keep in mind that body composition can differ from one person to another, which is why I like to focus more on waist size than weight.

The Ongoing Benefits of Exercise and Sleep

If you aren't working out on a regular basis, now may be a good time to start. Exercise can not only improve your ability to stay in a state of nutritional ketosis, it also has many other significant health benefits, including greater bone density and musculature, increased production of the hormones that improve mood, and better brain health. That's why it's important that, at a minimum, you continue to go for walks. However, you may also discover in Step IV that, as the weeks progress and your weight decreases, your energy levels and stamina increase. This presents a good opportunity to include more vigorous exercise in your weekly routine, under the supervision of your physician.

A consistent aerobic exercise program can help reduce high blood pressure, strengthen your immune system, and speed the fat-loss process. Anaerobic exercise, including weight lifting and resistance training, helps protect your joints, builds and maintains lean muscle mass, and boosts metabolism. Start with two or three sessions per week and slowly work your way up in intensity and frequency. Your body and brain will reap the benefits.

Anyone who's ever slept like a log after a full day of hiking knows that exercise also improves sleep, which, in turn, plays a part in weight loss. As we've learned, not getting enough sleep raises blood-sugar levels, and that shuts down ketosis. It also disrupts normal hormone levels, and that can interfere with metabolism and fat mobilization. So, turn off the TV, leave your phone in another room, and, if you can, get a minimum seven hours of sack time. Not only will it help you lose weight, but you'll also look well rested. Beauty sleep, indeed.

Measuring Success: The New You

By the end of Step IV, you have been on the BioDiet journey for 10 to 12 weeks. You have converted your metabolism from burning sugar to burning fat, lost weight and inches, gained energy, and improved sleep. Now's the perfect time to revisit those measurements you've been taking for the last few months. I also recommend that you and your physician discuss redoing your blood tests. These, along with the measurements, will provide both of you with a complete picture of how well the BioDiet is working. My hope is that it's working so well that you decide to continue with it. This is particularly important if you still have weight to lose or want to maintain and build upon the health benefits you've experienced so far.

I talk more about that in the next and final step of the BioDiet, Bio-Continuation, as we look at some of the options available to you, including intermittent fasting and transitioning to a low-carbohydrate lifestyle. Before we do that, however, I want to truly congratulate you for a job well done and encourage you to make the most of whatever well-earned reward you promised yourself for completing this step. You've earned it!

9

Step V: Bio-Continuation
What's Next on the Agenda?

*Success is a science; if you have
the conditions, you get the result.*
**OSCAR WILDE, AUTHOR, POET,
AND PLAYWRIGHT, 1854–1900**

BEGAN THIS BOOK by talking about science and the important role it plays in determining what we eat: why the foods we were told could kill us actually help us achieve good health, while the foods we were encouraged to eat can, in fact, lead us to an early grave. As a scientist, coming to this understanding was pivotal, because these conclusions fly in the face of something I—and pretty much everyone else—had believed for decades: that by replacing high-fat foods with those rich in carbohydrates we were doing the best thing possible for our bodies and brains.

We now know that isn't true, and the results of this massive experiment on the trusting North American public have, instead, produced the Axis of Illness—a deadly trifecta of obesity, insulin resistance, and systemic inflammation—along with the resulting host of chronic diseases, including cardiovascular disease, cancer, diabetes, and Alzheimer's. It's time to address this growing health crisis by adopting a new way of eating that reverses weight gain and obesity, increases insulin sensitivity, and decreases the chronic inflammation that is at the root of so many debilitating and deadly diseases.

So, despite the fact the many health professionals still adhere to a dogma that has never been scientifically validated and has failed the test of time, I, along with many others, work to upend the conventional "wisdom." If that makes me a heretic, so be it. What about you?

Over the past 12 weeks on the BioDiet, you reset your metabolism, improved your vitality, reduced your risk of chronic disease, and likely lost much, if not all, of that spare tire you may have been toting around for ages. And those are just the physiological benefits. Psychologically, you took great strides by recognizing the need to make positive changes in your life: you committed to a plan that would help you accomplish those changes and then followed through. You gave up what may have been some of your favorite foods in exchange for becoming leaner and healthier; you took responsibility for your own nutrition and changed your relationship with food for the better; and you got a little exercise, perhaps even embarking on your first regular fitness regimen. I hope you are very proud of all that you've accomplished. You should be! You are now well on your way to a healthier, happier life.

Bio-Continuation: Overview

I like to think of this step as a time to enjoy your newfound health and improved physique and to give thought as to where you want to go from here. That decision depends on whether you've achieved your goals for lost weight and inches, improved your health sufficiently, are happy to continue with the dietary changes you've made, or want to consider other options. Here, we explore these questions along with some other factors to consider when making what will be an important decision for your future.

How long does Step V last?

That's up to you, though I hope that after weighing the benefits of the BioDiet, you'll stay with it for life.

What changes will I be making in my diet?

That depends on whether you want to shed additional fat or simply maintain your new weight and waistline. I give you diet options for both. If you've adopted a regular fitness routine, I also have a few suggestions for what to eat before and after exercise.

What do I need to buy?

I'm hoping you've rewarded yourself with some new clothes to fit your new, slimmed-down physique. Other than that, no purchases required.

What else do I need to do?

If you haven't already, revisit your physician and consider having your blood work done again. This allows both of you to compare the results to those you received prior to embarking on the BioDiet. It is very likely that your blood pressure, blood sugar, and blood tri-glycerides have all improved. Consult with your doctor about your options going forward.

What physical and psychological changes can I expect?

If you remain in nutritional ketosis, you can, if you choose, continue to lose weight, though perhaps more slowly than before. I provide suggestions on how to rekindle the process if your weight loss has stalled. Keep in mind, though, that in terms of health, a better measure than weight loss is waist size. This, too, has decreased over the last 12 weeks and should continue to do so.

In terms of psychological changes, these vary from person to person, but those I've counseled report improved mood and brain function and greater mental and emotional hardiness. As always, take note of any changes you experience on the BioDiet. These notes might be helpful as you consider what to do next.

Decision Time: BioDiet Options

Seven years ago, when my wife and I started on the BioDiet, I knew that, in theory, it should result in fat loss and improved health indicators. What I didn't anticipate was how much better we'd feel overall. By eliminating carbohydrates, we'd reduced inflammation, so we had less joint and muscle pain, and we'd stabilized our blood sugar, thus improving mood. We weren't hungry all the time, and we had a lot more energy. Sure, there were things we missed (did I mention I really like beer?), but the benefits of the BioDiet greatly eclipsed any downsides. By the time we'd reached our goals for losing weight and girth, the decision to stay on the BioDiet was a no-brainer. That's when I fundamentally understood that this was not a "diet" in the common sense of the word; it was a lifestyle that offered not just weight loss but also tangible health benefits. For these reasons, I hope to convince you to embrace the BioDiet for life, but I'm aware that not everyone will want to or be able to do that.

With that in mind, I'm going to present a few scenarios that might help you determine what Bio-Continuation will look like for you. By referring back to your notes from these past 12 weeks, you'll be more cognizant of the positive changes that occurred as a result

Figure 9.1: Decision time.

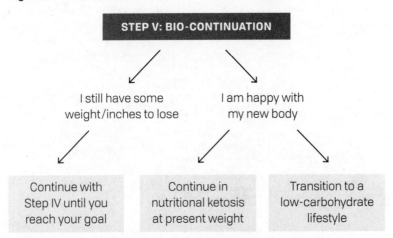

of the BioDiet, as well the challenges you may have experienced along the way.

As you can see on figure 9.1, there are three basic options to consider in moving forward, and returning to a diet rich in carbohydrates is not among them. That's because I really hope the one thing you take from this book is that the high-carbohydrate diet many of us have been eating all our lives is making us sick. So, let's explore some of the healthy alternatives.

You Still Have Weight/Inches to Lose

After three months on the BioDiet, men typically lose 20 to 25 pounds and women 10 to 15 pounds. This difference is due in part to the fact that men have higher metabolic rates than women, which means they burn more calories per hour. They also accumulate more visceral fat, which is the first fat mobilized for energy on the BioDiet. Women have more subcutaneous fat, which is harder to lose than the visceral kind, and tends to accumulate on the hips and

After three months on the BioDiet, men typically lose 20 to 25 pounds and women 10 to 15 pounds.

buttocks rather than in the midsection. That's actually a good thing, since fat deposits there don't correlate strongly with increased risk to health.

If, at this stage, you're still carrying more weight than is healthy, don't be discouraged. The rate at which we shed weight and inches varies from person to person. Rather than focus on how much weight you lose each week, look instead at your monthly numbers. And keep in mind that weight loss *should* be gradual. If you're losing too much too quickly, there's a good chance that you're calorie restricting, which we know can lead to *increased* weight and serious health problems in the long run.

By now, you should have a pretty good idea of your carbohydrate sensitivity. If adding higher-carbohydrate foods in Step IV caused weight gain and/or increased joint and muscle soreness, return to

Step III. This should get you back on the road to weight loss within a few days and resolve any issues associated with inflammation. In fact, I advise those I counsel to return to Step III once or twice per year with or without weight gain. This ensures that they're in a continuous state of ketosis and, if not, helps restart it.

You're Happy with Your Weight but Want to Remain in Nutritional Ketosis

Even if you've achieved your weight-loss goal, remaining in nutritional ketosis has many health benefits. Not only does it ensure that you don't regain the weight you worked so hard to lose, but it's also your best hedge against chronic disease. In order to stay in nutritional ketosis, however, you need to be on the lookout for carbohydrate creep. Watch out for hidden carbohydrates—in the form of fillers—found in many prepared foods, such as veggie burgers, crab cakes, and meatballs. Avoid anything that includes breadcrumbs, which can have between 14 and 20 grams of carbohydrates per quarter cup. Gravies can also contain a boatload of carbs, particularly those thickened with flour or cornstarch. Sauces and condiments, especially those served with fast foods, can also contain their fair share of carbohydrates, especially sugar. And remember gluten-free does not mean low-carb! Many gluten-free products not only contain more sugar than the gluten-containing products they're replacing, but they're also often made with high glycemic index gluten substitutes such as rice flour, arrowroot, and tapioca starch, which can send your blood glucose skyrocketing.

You Want to Transition to a Low-Carbohydrate Diet

If you don't think you can sustain the lifestyle you experienced in Step IV, or simply don't want to, then it's worth experimenting with a regimen that's low-carbohydrate but not ketogenic. This means that you won't reap the additional health-promoting benefits of ketones, but you should be able to maintain some of the other advantages of low-carbohydrate consumption, though to a lesser degree.

Achieving a balance of low-carb/moderate-fat intake can be a challenge, and too much of either can cause you to gain weight, re-spark inflammation, and increase insulin levels. To help prevent these issues, start by adding to your diet vegetables that have some starch but lots of fiber; this slows down the absorption of glucose into the blood. By choosing lower glycemic index fruits, such as apples, pears, and kiwis, you can also limit the impact on your blood-sugar levels. You'll find a list of common vegetables and fruits, along with their carbohydrate load and glycemic index, in appendix D.

It's very important that you avoid foods that contain both high glycemic index carbohydrates and saturated fats—things like ice cream, baked goods, and pizza. While saturated fats on their own pose no health risk, when they are consumed with high glycemic index carbohydrates, they will elevate blood triglycerides. That's because your body will burn the glucose first, leaving the fats to circulate in the bloodstream and eventually find their way to the unhealthy adipose tissue you have been working so hard to reduce.

The sudden reintroduction of carbohydrates to your system can also result in some uncomfortable side effects, including intestinal pain, bloating, gas, and diarrhea or constipation. You can minimize these symptoms by adding carbohydrates to your diet gradually, which allows your body to slowly adjust. However, repeatedly bouncing back and forth between nutritional ketosis and a low-carbohydrate, non-ketogenic diet isn't advised. The constant shifting of your metabolism can result in the chronic inflammation, elevated insulin levels, and weight gain that got you into trouble in the first place.

Yo-Yo Dieting

I'm sure many of you are familiar with this frustrating routine: you go on the latest fad diet because you want to lose weight, but the moment you go off it, you regain what you lost and, oftentimes, more. So, you start a new diet only to have the same thing happen all over again. There are two main reasons we gain weight after a period of dieting. The first is that after weeks or months of calorie restriction, we're often nutrient deficient and hungry, so we tend to eat more than we might otherwise. But there is also a genetic reason we pack on the pounds after dieting, and it has to do with the cycles of feast and famine experienced by our hunter-gatherer ancestors.

When we calorie-restrict, we activate genes that evolved to regulate efficient intake and use of food stores. These "thrifty" genes help us conserve energy by lowering our metabolism when food is scarce (famine) and return it to normal when food becomes plentiful again (feast). However, the rise in metabolism doesn't happen immediately and, in the interim, we tend to hungrily consume more calories than we need. The result is weight gain. What's more, those thrifty genes prepare our bodies for future famine by storing calories as fat. This is why when we return to our regular diet, we gain back the fat we lost and then some.

The weight fluctuations associated with yo-yo dieting not only take a toll on our mental well-being, by oftentimes giving rise to feelings of failure and even shame, but they can also have a serious impact on our physical health by causing swings in blood pressure, cholesterol, and blood sugar. A 2018 study that looked at the data from 6.7 million people over seven years found that those whose weight fluctuated most dramatically between physician appointments were 43 percent more likely to suffer a heart attack during the study period, 41 percent more likely to have a stroke, and a shocking 127 percent more likely to die. Be kind to your body by choosing a way of eating that benefits you in the long term. Your life might depend on it.

Exercise and the BioDiet: The Ongoing Benefits

Throughout the previous chapters, I discussed exercise as it relates to each of the BioDiet steps, but I think it's important to explore, in a more general way, the impact of nutritional ketosis on exercise and vice versa.

Whether you work out regularly or have only recently taken up exercising, you may find that on a ketogenic diet, you feel less

Intermittent Fasting: Latest Trend or Long-Term Lifestyle?

Throughout recorded history, there are many references to fasts being used for both health and spiritual benefit. Paracelsus, one of the fathers of Western medicine, described the practice as "the greatest remedy— the physician within." Fasting results in a drop in blood glucose, which, as we know from chapter 4, triggers gluconeogenesis and the production of beneficial ketones. Recent research has investigated short-term fasting as a means of treating disease, particularly those affecting the brain, such as Alzheimer's disease. Fasting has also been shown to be an effective means of weight loss, and some studies suggest it can reduce inflammation and may have a positive impact on insulin resistance.

Recently, the notion of *intermittent* fasting has become popular. Rather than going without food for extended periods, which can have serious health implications, intermittent fasting prescribes eating normally for a set period of time and then significantly restricting your food intake. The 16:8 method, for example, involves a daily "eating window" of eight hours, whereas the 5:2 diet allows you to eat normally five days of energy at the start of your workout than you'd like. To explain why, I'd like to return to our cabin analogy from chapter 2. When we use kindling—carbohydrates—to fuel our cabin, it produces energy quickly, but it burns fast. The same goes for exercising on a high-carbohydrate diet; we start strong but risk running out of steam midway through our workout. This is why you see many athletes refueling during an event or game. A diet high in fat, however, is like a log on the fire; it might take a little longer to get going, but once it does, it burns steadily over a much longer period.

the week and then requires that you restrict your food intake to between 500 and 600 calories the other two days.

The long-term effects of intermittent fasts are unknown, and there are some people for whom fasting is not recommended, including those who are underweight or have a history of eating disorders. If you suffer from diabetes, poor blood-sugar regulation, have low blood pressure, or are pregnant or trying to conceive, I strongly recommend you speak with your physician before you consider adopting a fast of any sort.

Since the general principle behind intermittent fasting is that it's *when* you eat not *what* you eat that matters, it doesn't necessarily consider carbohydrate-rich foods or those with a high glycemic index. This worries me, because these foods can have an impact on your health no matter what day of the week or time of day you consume them. If you are going to adopt intermittent fasting, I'd suggest you do it gradually and monitor not just weight loss but also any impact on your energy level, mood, brain function, and sleep.

When it comes to exercise, rather than hit the ground running, quite literally, take time to warm up. This will allow your fat-adapted muscle cells to adjust to their new, more efficient way of converting fuel to power. This is especially important if you're planning a vigorous cardio workout or high-intensity interval training.

As a general rule, building muscle requires the consumption of more calories than your body burns, which is why longer, harder, and more frequent workouts require more fuel. On the BioDiet, that means eating more fat and protein. By consuming enough of both, you ensure that your body doesn't begin to break down important muscle proteins as a source of energy.

Speaking of muscle, you *can* gain muscle mass on a ketogenic diet despite what some in the weightlifting world might have you believe. I estimate that I added about 10 to 15 pounds of muscle to my upper body—arms, shoulders, chest, and back—in the first six months or so on the BioDiet (and I'm in my 50s). If, however, your goal is to look like Serena Williams or The Rock, you'll need to spend a lot of time and effort pumping iron. I advocate for a more well-rounded approach to exercise, one that includes not just weightlifting but also aerobic exercise, flexibility, and stability training. The older you are, the more important all of these become.

I am often asked whether it's better to eat before or after exercise. There are many schools of thought on this, but in general you want to work out a little hungry and then eat soon after to maximize the uptake and incorporation of protein into those hard-working muscles. However, if you are going to engage in high-intensity sports, like basketball or hockey, you'll probably perform better with some carbohydrates in your system. Fifty grams or so, 20 minutes prior to start time should supply your body with enough fuel to ignite your muscle metabolism, at which point the fats take over to provide a steady and efficient source of energy. For longer events that have sustained high output, like cycling races or triathlons, your active muscles will burn through those carbohydrates, which is why, according to anecdotal reports, some athletes can consume as much as 150 grams of carbohydrates per day and still maintain nutritional ketosis.

As anyone who's ever been sore after working out knows, exercise can take its toll on your muscles. That's where sleep comes in, particularly the first 90 minutes of sleep. As you'll recall, your first sleep cycle is when your body releases the majority of its growth hormone, which stimulates protein synthesis to build, maintain, and repair tissues. You can optimize this process by ensuring your body has sufficient amino acids during your first sleep cycle. The best way to do that is to consume protein (which you'll recall from chapter 2 breaks down into amino acids) before turning in. Although this is contrary to my recommendation to "refrain from eating anything two hours before going to bed," it is an exception to the rule for those looking to build muscle. To optimize growth hormone, it is important that there are no carbohydrates in what you consume. I like a shake made with whey and casein protein powders. Whey addresses the short-term needs, especially during your first sleep cycle, and the amino acids from casein work more slowly throughout the night.

Frequently Asked Questions

I'm fairly sure that, despite my best efforts to be thorough, you will still have questions about the BioDiet. That's great, because it means you're taking your health seriously. While you'll find answers to many of your queries on BioDiet.org, I thought I'd take the opportunity here to briefly answer those questions I'm most frequently asked.

1. *I am a vegetarian. Can I adopt the BioDiet?*

Yes! You don't have to be a carnivore to be a BioDieter. The Bio-Diet is high in fat, not high in protein. Therefore, as long as you're getting enough healthy fats from sources like avocados, nuts and nut butters, olives and olive oil, and sufficient protein and essential nutrients from soy, eggs, and dairy (which also contain beneficial fats), broccoli, spinach, asparagus, and brussels sprouts, for example, you can still enjoy a vegetarian diet while getting all the benefits of the BioDiet.

2. *I am vegan. Can I adopt the BioDiet?*

Yes, you can, but you need to ensure that you get complete proteins containing all nine essential amino acids. Since you won't be eating any animal products, these amino acids will need to come from soy (tofu, tempeh, and edamame) and seeds (pumpkin, chia, and hemp seeds). However, keep in mind that, because soy mimics the hormone estrogen, it could interfere with a woman's menstrual cycle and her ability to become pregnant. Also, some important nutrients, including B_{12} and folic acid, are not found in plants, so you need to take those as supplements.

3. *Will my cholesterol rise on a ketogenic diet?*

The short answer is yes, for some, but not for everyone. And if it does rise, that might not be a bad thing. As mentioned throughout the book, there are two general categories of cholesterol in the blood. Low-density lipoproteins, or LDL, are considered to be the "bad" form. LDL contributes to the atherosclerotic plaques that build up on arteries and lead to cardiovascular disease, which can manifest as heart attack or stroke. On the other hand, high-density lipoproteins, or HDL, are considered to be "good," because they do the opposite, scavenging cholesterol from the plaques and transporting it to the liver where it can be safely metabolized or stored.

This reduces the risk of cardiovascular disease, which is why rising HDL is seen as beneficial. What's more, higher levels of HDL are also associated with decreased inflammation, another positive outcome.

However, these assignments as "good" or "bad" are an oversimplification. LDL is not bad: if your LDL was zero, you'd be dead. We need LDL to get cholesterol from source—the liver—to the rest of the cells in the body. Cholesterol is an integral part of cell membranes, an important component of all cells; all of your steroid hormones are synthesized from cholesterol; and it is needed for the synthesis of vitamin D and bile salts, the latter being necessary for the absorption of fats and fat-soluble vitamins.

LDL correlates with increased risk of cardiovascular disease. However, the correlation is not as strong as you might think. In fact, half of people who have heart attacks and strokes have absolutely normal LDL. How can this be? The answer is that there are actually three types of LDL: low-density (just called LDL), intermediate-density (IDL), and very low-density (VLDL), which is the light, fluffy variety. Only the small, particle-sized LDL correlates with increased risk of cardiovascular disease. So, you can get a rise in overall LDL cholesterol, but that could be due to increased IDL and VLDL, which don't correlate with increased risk.

Unfortunately, standard blood lipid panels don't measure the different particle sizes. In fact, they don't measure LDL at all. What they do is estimate LDL-c (c for concentration) from an equation that involves measuring blood concentrations of total cholesterol, HDL, and triglycerides. Many clinicians and physicians are now looking to other biomarkers that more effectively predict the risk for cardiovascular disease. One that can be discerned from the standard lipid panel is the ratio of total cholesterol to HDL, which is a better predictor than just LDL. Another is the ratio of triglycerides to HDL—the lower, the better—and it is probably better still as a predictor. It's also important to note that a very small proportion of people will experience a significant, rapid rise in triglycerides on the BioDiet. If this happens, please follow your physician's advice, which will likely be to discontinue.

There are three ways to raise HDL: consumption of saturated fat, moderate consumption of alcohol, and exercise. These are all elements of the BioDiet, so you should expect to see a rise, perhaps as much as a doubling, of your HDL on the BioDiet. This will, of course, also result in a rise in total cholesterol.

4. *Can I afford the BioDiet?*

Yes. One of the best things about a ketogenic diet—other than the health benefits, of course—is that trips to the grocery store don't involve carrying home mountains of food. That's because, unlike high-carbohydrate diets rich in glucose and starch, the foods you consume on the BioDiet are so nutrient-dense that you don't need to eat much to get the vitamins and minerals your body needs. Because the BioDiet relies largely on fresh foods, you actually consume the vegetables you purchase rather than later discover them rotting in the back of the refrigerator.

There is no doubt that grains and grain-based foods can be mass-produced at low cost, but fats can also be cost effective. For the sake of argument, let's say you decided to get all your fat calories from high-quality, extra-virgin olive oil. You'd need to consume about 150 grams of it per day, or about two-thirds of a cup, to get all of the fat calories you need. That could cost as little about $1.50 a day.

5. *What would happen to the environment if the majority of people adopted the BioDiet?*

This is a question I'm often asked and it's a difficult one to answer, because there are many aspects to consider. What I can say is that, because the BioDiet relies largely on fresh food, it requires less

There are a quarter of a billion obese and overweight people in the United States and Canada, and I believe that by changing their diet, they can significantly improve their health outcomes, saving our already overburdened healthcare systems hundreds of billions of dollars.

manufacturing and, therefore, less energy use and packaging waste. And, because it's high fat and fat is twice as energy dense as carbohydrate, you consume less than half the food weight compared to the standard Western diet. This significantly reduces the environmental costs associated with shipping. Vegetarians who adopt the BioDiet can reap its health benefits, while at the same time reducing the need for animal agriculture.

Before I leave this question, I'd like put forth another reason for adopting the BioDiet, or any other ketogenic lifestyle, for that matter. There are a quarter of a billion obese and overweight people in the United States and Canada, and I believe that by changing their diet, they can significantly improve their health outcomes, saving our already overburdened healthcare systems hundreds of billions of dollars, and providing our cash-strapped farmers with the financial incentive to grow nutritious, environmentally sustainable, BioDiet foods.

Some Final Thoughts

I hope this book has been as helpful for you to read as it has been for me to write. It has required me to think carefully about the importance of food, not just in terms of our individual health but also the value it can bring to our society as a whole. I can't help but imagine a future when, despite our aging population, preventable chronic diseases are on the decline and obesity is a rarity rather than the status quo. I look forward to a time when we can wrestle back control of what we eat from processed-food manufacturers that have little concern for our health and well-being and a great deal invested in their bottom lines.

You have learned more about your own human biology, and that empowers you to keep your body healthy. Of course, with that power also comes the responsibility to decide what works best for you. I want to thank you for giving me your time and attention, and I look forward to meeting as many of you as possible in person. In the meantime, I will continue to advocate for evidence-based public policy on nutritional recommendations and work to improve our collective health.

On BioDiet.org, you'll find research updates, recipes, deeper dives into the science, and reports on efforts to change nutritional public policy. You'll also have a means to connect with others in the growing keto community to share your experiences and insights.

At the start of our journey, I asked you to commit 12 weeks of your life to the BioDiet. I hope you're now feeling healthier and happier—and looking great. That has been my goal throughout, and it has been a privilege to be your guide. BioDiet for life!

Acknowledgments

THANK MY WIFE, Dale, for joining me on the BioDiet and partnering with me on this book. I could not have done this without her. The wonderful, patient, and very supportive team at Page Two provided the logistics needed to produce a book, and our editor, Linda Pruessen, was invaluable helping with structure, tone, and focus. Copy editor Crissy Calhoun provided great advice, and I also thank Judy Kelley for her editorial work and helpful comments. It was Dr. Richard Mathias who first took me down the rabbit hole and pointed me to the science behind the BioDiet, and Drs. Jeff Volek and Stephen Phinney who greeted me upon my arrival, providing inspiration and guidance. I am very grateful for their continued generosity of time and knowledge. I have learned so much about immunology and cancer biology from my colleague and good friend Dr. Gerald Krystal, and I thank him for welcoming me into his lab at the BC Cancer Research Centre, Terry Fox Laboratory. It has been a privilege as well as a great personal and professional experience to work with the extraordinarily bright and talented young people in Jeff's and Gerry's labs. To my family, friends, and all those I've met over the years who have heard my thoughts about nutrition and provided critical feedback, this has been integral to my success creating this work.

Appendix A
BioDiet Grocery List

ERE ARE THE core foods for the BioDiet. Since Step III is the most restrictive, I have indicated the foods you can't eat until Step IV with an asterisk (*). If you want to add a food to your diet that is not on the list, please check its nutritional content on one of the many websites with that information. (You can find those resources on BioDiet.org.) In general, apply these simple rules to determine if a food is BioDiet friendly.

1. Processed foods are very unlikely to be BioDiet friendly unless they are well formulated for ketogenic diets. This includes much of what you'll find in the center of the grocery store, so stick to the outer edges when shopping.

2. Animal products are generally BioDiet friendly, but keep an eye out for high-carbohydrate fillers in foods such as sausages or added sugar in products like yogurt. Bacon also typically has considerable added sugar, so watch for that.

3. Some vegetables, such as root vegetables, should be avoided. Eat vegetables that grow above the ground that are not beans (green and yellow wax beans are fine) or grains (including quinoa).

4. Fruits, other than berries, should be eaten sparingly. Avoid bananas, which contain a lot of starch, and all fruits with a high glycemic index.

Vegetables and Fruits

Leafy Greens
- Cruciferous vegetables (kale, collard greens, watercress, broccoli, brussels sprouts, kohlrabi, cauliflower, bok choy, rapini (broccoli rabe), cabbage, arugula, radish)
- Lettuces (especially romaine)
- Parsley
- Spinach

Other Common Vegetables
- Artichokes
- Asparagus
- Avocados
- Celery
- Cucumber
- Eggplant*
- Endive
- Mushrooms
- Onions (not sweet)
- Peppers (all types)
- Radicchio

- Snow peas*
- Summer squashes*
- Tomatoes
- Wax beans (green and yellow)
- Zucchini

Fruits
- Lemons and limes (used sparingly for flavoring teas and water)
- Olives
- Berries* (blackberries, raspberries, strawberries, blueberries)
- Apples* (very sparingly)
- Kiwi* (very sparingly and eat the rind)
- Pears* (very sparingly)

Proteins

(Buy organic products that contain ample fat and, if comes with skin, eat that, too.)
- Eggs (there's one on the cover of the book, right?)
- Fish (especially salmon, tuna, sardines)
- Shellfish
- Poultry (chicken, turkey, duck, goose)
- Pork (bacon, ham—especially prosciutto or serrano)
- Beef
- Lamb
- Sausage (any kind but watch for high-carbohydrate fillers; try to get all-meat varieties)
- Tofu, tempeh, and textured vegetable protein (especially for vegetarians)
- Non-wheat, non-grain flours* (almond, coconut, soy)

Fats and Oils

- Butter (especially grass-fed butter)
- Olive oil (only extra-virgin)
- Macadamia oil
- Avocado oil
- Coconut oil
- Beef tallow

Dairy

- Cheeses* (especially very high-fat cheeses (>30% fat: cheddar, Swiss, and gouda) during Step III, only 2 oz/day (¼ cup); after Step III you can also include high-fat cheeses (>20% fat) such as Parmesan, Brie or Camembert, blue, Jarlsberg, goat cheese, feta, and mozzarella)
- Yogurt (plain, unsweetened; I recommend high-fat Greek style)
- Whipping cream; full-fat sour cream; full-fat cream cheese
- Kefir (unsweetened)

Nuts and Seeds*

(During Step III, only 2 oz/day (¼ cup) total.)
- Chia seeds (whole or ground; important fiber source)
- Flax seeds (important fiber and omega-3 fatty acids)
- Hemp seeds
- Pumpkin seeds
- Almonds (including almond flour and almond butter)
- Brazil nuts (go to these first when hungry between meals)

- Hazelnuts
- Macadamia nuts
- Pecans
- Pistachios (higher in carbs, so eat sparingly)
- Walnuts

Between Meals

- Olives
- Pickles (made with no sugar)
- Pork rinds (chicharróns)
- Beef jerky
- Sugar-free chocolate (small amounts, 1 oz/day maximum)
- Sugar-free candies (small amounts, 1–2/day maximum)

Beverages

- Water (fresh-filtered and/or carbonated)
- Coffee (with full-fat or whipping cream)
- Tea (especially black and green; other teas are fine as well; no bottled teas)
- Nut milks (soy, almond, coconut; unsweetened)
- Clear broth (full salt during Bio-Adaptation step)
- Wine*
- Spirits, low-carbohydrate beer*

Flavorings

- Salt (iodized)
- Herbs and spices (all are fine, but the following have anti-inflammatory or other health benefits):
- Basil
- Black pepper
- Cardamom
- Cayenne pepper
- Chives
- Cilantro (coriander)
- Cinnamon
- Cloves
- Garlic
- Ginger
- Nutmeg
- Oregano
- Parsley
- Sage
- Rosemary
- Turmeric
- Vinegars, Worcestershire sauce
- Mustards (unsweetened)
- Hot sauces (unsweetened)
- Soy sauce
- Extracts (vanilla, almond, maple)
- Sugar substitutes (e.g., xylitol or erythritol, used sparingly)

Appendix B
Sweeteners

D ESPITE SUGAR'S INSIDIOUS health consequences, it can be a hard habit to kick. That's why science and industry have come up with a variety of sugar substitutes, including artificial sweeteners, sugar alcohols, plant-based sweeteners, and alternative sugars. Here's a cheat sheet to all your choices.

Artificial Sweeteners

Artificial sweeteners—such as sucralose (Splenda), aspartame (Equal, NutraSweet), acesulfame potassium (Sweet One, Sunett), and saccharin or cyclamate (SugarTwin, Sweet'N Low)—contain virtually no calories and, because they are generally hundreds of times sweeter than sugar, a little goes a long way. They are widely used in processed foods but also popular for home use. All are generally considered safe for consumption in limited quantities.

For those, myself included, who find artificial sweeteners to have an unpleasant aftertaste, *sugar alcohols* may provide an alternative. Sugar alcohols are, in fact, neither sugars nor alcohols, but a class of organic compound called polyols and include xylitol, erythritol, and maltitol.

- *Xylitol* (Xyla) is perhaps the best known of the bunch. Used as a sweetener for more than a century (check your sugarless gum, Tic Tacs, or mouthwash—you may have been consuming the stuff for years), xylitol is derived from berries, mushrooms, corn-husks, and, in North America, hardwood trees, especially birch. It looks and tastes very similar to sugar but contains less than two-thirds of the calories and has a much lower glycemic index. And, like all sugar alcohols, xylitol isn't metabolized by bacteria in the mouth, so it doesn't contribute to tooth decay. However, because it's fermented by bacteria in the gut, xylitol can have a laxative effect and cause gas and bloating in some people. If you're going to try xylitol, I advise you to start slow. Also, keep in mind that, like chocolate, xylitol is poisonous to dogs because they lack the enzyme to metabolize it.

- If you do have problems adjusting to xylitol, you may want to try *erythritol* (Swerve), which does not ferment in your gut and has almost no calories and a very low glycemic index. Erythritol isn't as sweet as sugar, so you need to use about one-third more to achieve the same level of sweetness. It also tends to have a menthol aftertaste, which makes it less appealing to some.

- The other commonly used polyol is *maltitol*, which is used as a sugar substitute in many processed products including chocolate, hard candy, and chewing gum. However, maltitol has more calo-ries and a higher glycemic index than most other sugar alcohols, so may not be the best choice if you suffer from diabetes.

Although *plant-based sweeteners* have gained a lot of attention over the last decade, some have been used as a sweetener for centuries. Stevia (Truvia, Pure Via, SweetLeaf), which comes from a family of plants that grow wild in tropical climates, was long used in Paraguay and Brazil before appearing in Europe in the early 1900s. Today it is sold around the world and can be found in a variety of products including sodas and sports drinks, baked goods, and yogurts. It is also available in liquid, granule, and powder forms. Some find stevia has a bitter, unpleasant aftertaste, and, like sugar alcohols, it can cause mild stomach upset, gas, and bloating.

Monk fruit (Sugarlike), a small gourd from Southeast Asia, has also been used as a sweetener for hundreds of years. Like stevia, it has no calories and no glycemic index. Although monk fruit tends to have a better aftertaste than stevia, it is often more expensive and can be difficult to find.

Alternative Sugars

No matter what the clever marketers say, these are sugars by another name and decidedly incompatible with the BioDiet. Alternative sugars include honey, maple syrup, molasses, fruit juice concentrate, and agave syrup and are promoted as healthy alternatives to table sugar. However, given that their vitamin and mineral content isn't significantly different from that of sugar, they offer little or no health advantage. Just like table sugar, your body breaks them down into glucose and fructose. Agave is the worst offender because it's the highest in fructose, but all alternative sugars can lead to weight gain and obesity, insulin resistance, diabetes, and chronic inflammation.

Appendix C
Nutrition Labels

THE OLD ADAGE applies: read the label. Thankfully, nutrition labels are now required for almost all packaged foods. But what do those labels mean and how do they apply to the BioDiet? For an example, let's look at a label for canned baked beans (see following page):

Start at the top with the serving size, which in this case is a reasonable one-half cup. However, some manufacturers state unrealistically small serving sizes to give the appearance that the product contains less sodium, for example, than you might want in your diet. That means the only way you can figure out how much saturated fat, cholesterol, sodium, et cetera, you'll be consuming is to measure the amount you plan to eat and calculate the actual numbers from there.

One of the great things about the BioDiet is that you don't have to count calories, so there's no need to concern yourself with them. Nor do you need to worry about the percent daily value (%DV) for macronutrients (fat, carbohydrate, and protein) because the %DV is based on a standard Western diet, not a ketogenic diet.

If you want to know the percentage of fat in a given product, there are two ways to calculate it, depending upon the label. Some

Nutrition Facts

Serving Size 1/2 cup (174 g)
Servings Per Container 16

Amount Per Serving

Calories 190	Calories from Fat 20

% Daily Value*

Total Fat 2g	3%
Saturated Fat 0g	0%
Trans Fat 0g	
Cholesterol 5mg	2%
Sodium 480mg	20%
Total Carbohydrate 34g	11%
Dietary Fiber 8g	32%
Sugars 13g	
Protein 9g	

Vitamin A 2%	•	**Vitamin C** 6%
Calcium 6%	•	**Iron** 15%

*Percent Daily Value are based on a 2,000 calorie diet. Your daily values maybe be higher or lower depending on your calorie needs:

	Calories:	2,000	2,500
Total Fat	Less than	65g	80g
Saturated Fat	Less than	20g	25g
Cholesterol	Less than	300mg	300mg
Sodium	Less than	2,400mg	2,400mg
Total Carbohydrate		300g	375g
Dietary Fiber		25g	30g

Calories per gram:
Fat 9 • Carbohydrate 4 • Protein 4

labels, like the one above, show calories from fat, which in the case of our canned baked beans is 20. So, to determine the percentage of fat in the beans, divide the calories from fat by the total calories, which in this example is 190 and then multiply by 100.

$$(\text{calories from fat} \div \text{total calories}) \times 100 = \text{fat percentage}$$
$$(20 \div 190) \times 100 = 10.5\%$$

If the calories from fat don't appear on the label (or you want to calculate the fat percentage yourself), the math is a little more involved. In this equation, you would multiply the total grams of fat, which in the case of our beans is 2, by 9 (the number of calories provided by each gram of fat), divide the result by the total calories, which in this instance is 190, and then multiply that result by 100.

$$(\text{grams of fat} \times 9) \times 100 = \text{fat percentage}$$
$$(2 \times 9) \div 190 \times 100 = 9.5\%$$

In this example, there is a slight difference between the two results. This is due to the rounding of numbers on the manufacturer's label.

Moving further down the Nutrition Facts list, I suggest you don't worry about the amount of cholesterol in the product (remember there is no relationship between the cholesterol in our food and our blood cholesterol) or about the sodium. That's because the BioDiet contains no processed food, which, as you'll recall, is the leading source of sodium in the standard Western diet. In fact, those on the BioDiet need to ensure they get *enough* salt in their diet.

Now, let's look at the total carbohydrate count, which in the case of our beans is 34 grams per half-cup serving. (Beans are not low-carbohydrate, despite what many people think.) From that number, you subtract the grams of fiber, in this case 8, giving you 26 grams of net carbohydrate. Of those, 13 grams are sugar, and the remaining 13 grams are starch.

$$\text{total carbohydrates} - \text{grams of fiber} = \text{net carbohydrates}$$
$$34 - 8 = 26$$
$$\text{net carbohydrates} - \text{grams sugar} = \text{grams of starch}$$
$$26 - 13 = 13$$

There are 9 grams of protein in each serving of beans, or roughly 19 percent of the total calories. Add that to the roughly 10 percent fat and that means about 70 percent of the calories in the beans are from carbohydrate. These baked beans are definitely *not* a BioDiet-friendly product!

Below the macronutrients on our Nutrition Facts list are the appreciable amounts of vitamins and minerals (other than sodium), listed as percent of daily value. The higher the percentages, the more nutrient-dense the product.

Appendix D
Carbohydrate Content and Glycemic Indices for Some Common Foods

ERE ARE SOME common vegetables and fruits listed by their net carbohydrate content (total carbohydrate minus fiber). They are ranked for comparison by net carbohydrate (grams) for equal serving sizes. For the fruits, however, the carbohydrate load is also given for a typical serving size because we typically eat the whole apple, for example, and not just 50 grams. The glycemic index for each is also provided. Those fruits and vegetables with lower net carbohydrate and glycemic index and higher fiber are healthier choices, so try to select items higher on the list. Please note that most vegetables that grow above ground do not have validated glycemic indices. The value of 32 is assigned as a best estimate for denser vegetables, and 15 for those, such as cucumber and sprouts, which have high water content.

The carbohydrate values listed are kindly supplied by my colleague Teryn Sapper, MS, RD, LD, a research dietician and senior research associate at The Ohio State University. She is part of the team I work with, investigating the therapeutic benefits of a ketogenic diet for women with metastatic breast cancer. The glycemic index values come primarily from the *International Table of Glycemic Index and Glycemic Load Values: 2002* published in the *American Journal of Clinical Nutrition*, among other sources.

Vegetables

(100 grams = ½ cup serving)

Name	Net carb (per 100 gms)	Total carb (per 100 gms)	Fiber (per 100 gms)	GI index
Alfalfa sprouts	0.2	2.1	1.9	15
Collard greens	1.1	3.4	2.3	32
Spinach	1.1	3.5	2.4	32
Romaine	1.2	3.3	2.1	32
Celery	1.4	3.0	1.6	32
Mustard greens	1.5	4.7	3.2	32
Asparagus	1.7	3.9	2.1	32
Lettuce	1.8	3.0	1.2	32
Radish	1.8	3.4	1.8	32
Avocado	2.0	8.5	6.5	50
Zucchini	2.1	3.1	1.0	15
Swiss chard	2.1	3.7	1.6	32
Mushroom	2.3	3.3	1.0	32
Tomato	2.7	3.9	1.2	38
Rhubarb	2.7	4.5	1.8	13

Name	Net carb (per 100 gms)	Total carb (per 100 gms)	Fiber (per 100 gms)	GI index
Green bell pepper	2.9	4.6	1.7	32
Eggplant	2.9	5.9	3.0	32
Cauliflower	3.0	5.0	2.0	32
Olives	3.0	6.0	3.0	0
Cucumber	3.1	3.6	0.5	15
Cabbage	3.3	5.8	2.5	45
Red bell pepper	3.9	6.0	2.1	32
Broccoli	4.0	6.6	2.6	32
Bean sprouts	4.1	5.9	1.8	15
Green beans	4.3	7.0	2.7	32
Okra	4.3	7.5	3.2	32
Turnip	4.6	6.4	1.8	73
Scallion	4.7	7.3	2.6	32
Spaghetti squash	5.1	6.5	1.4	20
Pumpkin	5.2	8.1	2.9	75
Carrot (boiled)	5.2	8.2	3.0	49
Kale	5.2	8.8	3.6	32
Brussels sprouts	5.2	9.0	3.8	32
Yellow bell pepper	5.4	6.3	0.9	32
Artichoke	5.9	10.5	4.6	32
Rutabaga	6.3	8.6	2.3	72
Carrot (raw)	6.8	9.6	2.8	47
Butternut squash	7.3	10.5	3.2	51

Name	Net carb (per 100 gms)	Total carb (per 100 gms)	Fiber (per 100 gms)	GI index
Celery root	7.4	9.2	1.8	35
Onion	7.6	9.3	1.7	15
Beet	8.0	10.0	2.0	64
Peas	10.1	15.6	5.5	32
Acorn squash	10.2	14.6	4.4	75
Lentils	12.2	20.1	7.9	29
Leek	12.4	14.2	1.8	32
Parsnip	13.1	18.0	4.9	52
Red potato (boiled)	14.2	15.9	1.7	57
Black beans	15.0	23.7	8.7	30
Navy beans	15.6	26.1	10.5	38
Kidney beans	16.4	22.8	6.4	28
Russet potato	16.8	18.1	1.3	85
Pinto beans	17.2	26.2	9.0	39
Sweet potato (boiled)	17.4	20.7	3.3	61
Sweet corn	18.6	21.0	2.4	54
Chickpeas	19.8	27.4	7.6	28

Fruits

(50 grams = ¼ cup)

Name	Net carb (per 50 gms)	Total carb (per 50 gms)	Fiber (per typical serving)	Net carb	GI index
Blackberries	2.1	4.8	2.7	½ cup = ~3	25
Raspberries	2.7	6.0	3.3	½ cup = ~3	32
Strawberries	2.8	3.8	1.0	½ cup = ~4	40
Watermelon	3.6	3.8	0.2	1 wedge = ~21	72
Cranberries (sweetened)	3.8	6.1	2.3	½ cup = ~4	62
Lemon (juice)	3.9	4.1	0.2	1 fruit = ~4	25
Peach	4.0	4.8	0.8	1 fruit = ~12	42
Cantaloupe	4.1	4.6	0.5	1 wedge = ~6	65
Papaya	4.1	5.0	0.9	1 cup = 11	59
Grapefruit	4.5	5.3	0.8	½ fruit = ~11	25
Honeydew	4.6	5.0	0.4	1 wedge = ~13	65
Lime (juice)	4.6	4.8	0.2	1 fruit = 3.5	24
Orange	4.7	5.9	1.2	1 fruit = ~12	42
Apricot	4.8	5.8	1.0	½ cup = ~7g	57
Plum	5.0	5.7	0.7	1 fruit = ~7	39
Sour cherries	5.3	6.1	0.8	½ cup = ~8	22
Apple	5.7	6.9	1.2	1 fruit = ~21	38
Clementine	5.8	6.8	1.0	1 fruit = ~8	70
Pineapple	5.9	6.6	0.7	1 cup = ~18	59
Blueberries	6.0	7.2	1.2	½ cup = ~9	54
Pear	6.0	7.6	1.6	1 fruit = ~20	38

Name	Net carb (per 50 gms)	Total carb (per 50 gms)	Fiber (per typical serving)	Net carb	GI index
Kiwi	6.4	8.2	1.6	1 fruit = ~9	53
Sweet cherries	7.0	8.0	1.0	½ cup = ~11	(high)
Pomegranate	7.0	8.3	1.3	½ cup = ~13	67
Figs	8.2	9.6	1.4	2 figs = ~16	61
Grapes	8.6	9.1	0.5	½ cup = ~13	46
Mango	8.6	9.6	1.0	½ cup = 25	51
Banana	10.1	11.4	1.3	1 fruit = ~24	52
Jackfruit	12.6	13.4	0.8	½ cup = 18.5	75
Plantain (cooked)	14.4	15.6	1.2	½ cup = ~22g	38
Dates	32.8	36.0	3.2	1 date = ~16g	103

Index

About the Authors

AS A HEALTH educator and cancer researcher, Dr. David G. Harper has studied the impact of diet on human health for many years. The culmination of that extensive work is the BioDiet, a ketogenic food regimen that he and his wife began in 2012. The significant weight loss and health improvements they experienced led Dr. Harper to counsel hundreds of people on the BioDiet, with similarly consistent and impressive results.

Dr. Harper is an associate professor of kinesiology at the University of the Fraser Valley and a visiting scientist at the BC Cancer Research Center, Terry Fox Laboratory. He holds a PhD from the University of British Columbia and completed a post-doctoral fellowship in comparative physiology at the University of Cambridge. He is on the scientific advisory board of the Canadian Clinicians for Therapeutic Nutrition and is a member of the Institute for Personalized Therapeutic Nutrition.

DALE DREWERY is an award-winning journalist and writer with a keen interest in science and human health.